Emergency Preparedness and Survival Guide

Prepared by the editors of

Backwoods Home Magazine

P.O. Box 712
Gold Beach, Oregon 97444
1-800-835-2418
www.backwoodshome.com

The articles that appear in this book were previously published in *Backwoods Home Magazine*.

ISBN: 0-9655203-9-0

Cover art by Don Childers

Manufactured in the United States of America

Published by *Backwoods Home Magazine*, Inc.
P.O. Box 712, Gold Beach, OR 97444
www.backwoodshome.com

Table of Contents

Introduction

During 1998 and 1999 many people stored food, water, and other emergency items to prepare for Y2K. During that time I told *Backwoods Home Magazine* s readers that Y2K ***did not*** pose a serious threat to them, but that economic downturn and our own government s oppression of our rights did. I was roundly criticized by many people back then for shirking my journalistic responsibility by not warning people about the dangers of Y2K. History has shown who was the responsible journalist, and many people were embarrassed at having engaged in a panic over an imaginary catastrophe.

Today people are once again preparing for possible calamity, this time at the hands of terrorists. And once again I have to say that it is economic recession and our own government s oppression of our rights that we have to fear most.

Yet the terrorist threat is far more real than the Y2K threat ever was.

Therefore we have compiled this book for readers who again want to prepare to take care of themselves and their families in the event of any future catastrophe. And we have tried to do so as we did prior to Y2K without resorting to the folly of misinformation and panic.

The following chapters are adapted from articles that previously appeared in *Backwoods Home Magazine*, and they represent the best preparedness information from our 12-year history of teaching people how to be self-reliant. The articles address food storage, including canning and cooking from food storage, storing water, maintaining a reliable electricity source, storing various fuels, medical kits, self-defense firearms, vehicle kits, and a survival garden.

The book is brief but fairly thorough, meant to be a quick read on the basics of preparedness. If you want very extensive, in-depth information, we recommend you also get our companion CD-ROM. It contains dozens of articles comprising hundreds of pages of preparedness information gleaned from the back issues of *BHM*. See the last page of this book for ordering information.

BHM s readers have been prepared right along. This book will help others become equally prepared.

Dave Duffy
Publisher
Backwoods Home Magazine

Chapter 1
Surviving emergencies

By Jackie Clay

Emergencies are a fact of life in this world. They've been occurring since the earliest days of man, and they'll periodically occur again, like it or not. But they are something to prepare for, not to fear. Emergencies come in all forms—economic slowdowns that seriously affect your income, natural and many-made calamities that threaten your safety, and sudden, temporary disruptions in any of society's critical systems. Here's a brief overview of what you can do to prepare for them.

Fuel storage

Store up fuel for heating, cooking, lighting, and your vehicle before you need it. Wood and other fuel can be stored a long time, will be cheaper before there is a great demand, and it will be easier to get. We have at least 12 cords of dry firewood in the lot, and only burn about one a year. It's nice to have...just in case. Likewise, a barrel of kerosene is good insurance against spending dark evenings. You may not be able to buy it during an emergency, but it's pretty cheap now.

Have alternative lighting, whether it is kerosene lamps, Coleman lamps, or propane wall lamps. Being scared and in the dark is not a good emotional combination. And be sure to have replacement wicks, globes, and other parts....just in case. We had one drop of condensation fall on a burning Coleman lamp, popping the globe instantly....and had to do without the lamp for a month, until we could get out of our snowed-in cabin for a replacement. Now we have replacements for each lamp type.

Store up some gasoline, where local regulations allow. With life-extenders, a barrel of gas can last a long time without getting too "old" to burn well. This gasoline can be used to travel or to cut wood to keep warm. (See Chapter 2 on storing LP gas, gasoline, diesel, and kerosene.)

Supplies to store

- 25 pounds laundry soap
- 12 28 oz. bottles dish soap
- 73 rolls toilet paper
- sanitary napkins in sufficient quantity
- 8 gallons bleach (used for sanitation as well as laundry)
- 12 bars hand soap
- 6 24 oz. bottles shampoo
- personal products, such as toothpaste, deodorant
- chainsaw oil and other items to keep things running
- pet foods
- livestock feed
- 55 gallons kerosene for lighting
- 25 gallons Coleman fuel or other lantern fuel

Food storage

Have enough food stored up for at least a year, preferably two. This seems basic, but I know very few people, especially those with good jobs and "normal" incomes, who have a full pantry. With today's disposable society, many people shop daily for what they will eat for the next day's meals. The pantry (if they even have one) shelves are empty or filled with junk.

We keep two years' food stored up, from wheat to home-canned foods, just in case. (See Chapters 3, 4, 5, and 6 on canning and food storage.)

Medical needs

Build up an extensive medical kit. We use a large field box, which has two levels and many compartments. Get as much medical training as you can, such as EMT classes. You might not want to become an EMT, but you can learn much that will help your family in an emergency. Buy a good book or two, such as the *Red Cross First Aid* manual and the *Merck Manual*. Then stock up that medical kit.

Remember that "normal" things, such as cold medicine, may not be readily available.

Both my husband, a CNA, and I, a veterinary field technician, have extensive medical experience, so we include in our medical kit sutures and suture needles, injectable local anesthetic, an IV and IV electrolytes, epinephrine for shock due to allergies or drug reaction, a blood pressure cuff and stethoscope, plaster casting material, and more. One never knows if, and when, a doctor will be available during an emergency. Of course, if a doctor *is* available, one should never attempt to treat their own or others' injuries or illnesses. (See Chapter 9 on medical kits for emergencies.)

Other high priorities

A radio and batteries are other must-haves. This keeps you in contact with what's going on in the world. There are several brands of solar powered radios on the market today for not a lot of money. Batteries must be replaced or recharged. We also have a weather radio to keep up on weather conditions which can affect one's life much more when times are tough. Listening to this 9-volt radio each morning keeps us in tune to Mother Nature. This book's publisher, Dave Duffy, has a satellite internet connection to keep in touch with the news, weather, and friends. He needs no phone. A small generator will give him the little bit of electricity he needs to run the computer.

Does your family have adequate warm, sturdy clothing? If not, this is a high priority. Buy good hiking boots, jeans, and clothing that is practical, long-lasting, and warm. If you never need 'em, they won't take up much space in a closet. If you do, they will become essential.

How about alternate transportation...not your car or pickup? Walking is fine, but doesn't cover many miles in a day.

Some alternatives are bicycles, ATVs, motorcycles, snowmobiles (in snow country), and horses. All take much less gas to run than a car or truck, and bicycles and horses take none at all. Remember you don't have to use a car or truck, even though you are used to it. The Amish and other plain people have always used horses and other non-motorized vehicles for travel, and in Europe and many other countries bicycles are the norm for a lot of folks.

A medical kit

A good first aid book
Thermometer
Daily prescription meds for all family members
Antibiotics
Ointments for the eye, fungus & cuts
Antidiarrheal medication
Pain and anti-inflammatory medication such as aspirin
Burn treatment, such as *Burn Free*
Iodine/Betadine
Alcohol
Oral electrolytes (for dehydration from fever, diarrhea, stress)
Cold remedies
Cough medicines
Cough drops/throat discs
Bandages
Gauze
Cotton
Surgical tape
Scissors
Hemostats
Tweezers
Needles to remove slivers
A dental kit to patch dentures, replace fillings, etc.)

You may choose to include more, depending on your medical experience and background.

A year's food supply for your family

This is a sample list for my family, which is a family of three. Your family needs may differ quite a bit, due to your meal preference. However, if you use this list as a base, you won't go hungry. It also allows for "company" meals. This is a realistic pantry supply to last a year comfortably. Remember to rotate your supplies, using the oldest first, replenishing as you use, in order to keep relatively fresh foodstocks. If you have a family of 4, increase the amount by 25%, a family of 6, by 50%, etc.

GRAINS

- 300 pounds of hard wheat or in combination with 150 pounds of wheat and 150 pounds of flour.
- 50 pounds of dry corn to grind for cornmeal
- 50 pounds of soft wheat
- 50 pounds white rice
- 50 pounds brown rice
- 50 pounds oatmeal
- 25 pounds of masa harina de maize (corn flour for tortillas and tamales)

PASTA

- 15 pounds spaghetti
- 6 pounds assorted noodles
- 6 pounds lasagna

LEGUMES

- 50 pounds of pinto beans
- 50 pounds of combined other beans, such as navy, kidney, etc.
- 20 pounds of split peas
- 20 pounds lentils

DAIRY

- 18 #10 cans dry milk or in combination with boxes of store-bought dry milk
- 2 #10 cans cheese powder
- 5 #10 cans dehydrated eggs
- 3 #10 cans butter or margarine

SUGAR

- 50 pounds white granulated sugar
- 10 pounds brown sugar
- 10 pounds powdered sugar

SHORTENING/OIL

- 10 3# cans shortening
- 5 48 fl. oz. bottles vegetable oil
- 2 16 fl. oz. bottles olive oil

SALT

- 10 pounds iodized table salt (for pickling & meat preservation as well as table use)

VEGETABLES

- 104 pints of green beans
- 104 pints of sweet corn
- 104 pints of carrots
- 104 quarts of tomatoes

- 104 pints of tomato sauce
- 104 half pints tomato paste
- 104 quarts of potatoes and/or 22 pounds instant potatoes
- 26 quarts of squash or pumpkin
- 26 pints beets
- 2 #10 cans dehydrated sweet corn
- 4 #10 cans dehydrated peas
- 1 #10 can dehydrated onions
- 2 #10 cans dehydrated broccoli

FRUITS

- 52 pints peaches
- 52 pints apple sauce
- 52 pints fruit cocktail
- 52 quarts apples (includes pies, etc.)
- 52 pints pears
- 104 pints misc. fruits
- 1 #10 can raisins
- 1 #10 can dehydrated strawberries
- 2 #10 cans dehydrated apple slices
- 2 #10 cans dehydrated banana slices

MEAT

- 52 pints lean beef/venison roast
- 52 pints chicken/turkey
- 52 pints ham/fish/misc.
- 52 cans tuna
- 52 cans Spam
- 52 pints home canned hamburger for tacos, casseroles, etc.
- 1 #10 can ea. TVP (textured vegetable protein), bacon, chicken)

SEEDS

A heavy selection of garden seeds to replenish your food supply, should the period of hard times last longer than a few months. Always opt for the worst and prepare ahead.

Most garden seeds last for years, if kept dry. One notable exception is onion seed, which should be replaced yearly.

MISCELLANEOUS

- 1 pound baking soda
- 3 pounds baking powder
- 1 pound dry yeast
- spices usually used
- 25 dozen canning jar lids, wide mouth & regular
- coffee, tea, powdered drink mixes in sufficient quantity
- A grain mill to grind grains
- An Amish or other "cooking with basics" cookbook or two
- 1 gallon inexpensive pancake syrup
- An assortment of "treats", such as pickles, jams, preserves

Stay-at-home emergencies

Some emergencies, such as severe ice storms, power outages, and blizzards occur every year or so in many cold-climate areas, more frequently in other areas. These create emergencies that largely demand people stay where they are out of common sense.

Remember, without power, you will not have access to stores for food and supplies, banks and ATMs for cash, a flush toilet, drinking water out of the tap, your furnace fan (or heat at all, if you have a "modern" all electric home), the electric kitchen stove, gas for your car, and many other "normal" conveniences you are used to.

The food in your refrigerator will slowly spoil and the food in your freezer will slowly thaw, then rot, if your electricity goes off for a long period. (See Chapter 8 on "What if the electricity goes off?")

Getting along comfortably without electricity

There's a lot you can do without electricity. Number one, begin a game plan of *not* depending on electricity.

Trade in your electric stove for a gas range that does not have electric ignition. It will work when there is no electricity. Better yet, buy a wood range for the kitchen or basement. Trade in your electric refrigerator for a gas unit. Not only will it run without power, but it keeps food much better and lasts for many more years. Have at least a back-up heating unit that does not depend on electricity to run or to power the fan. A wood stove, a propane or natural gas wall furnace, or a space heater are all more dependable than a central furnace requiring electricity to provide heat for the home. (A good book to buy is *Living Without Electricity and Liking It* by Anita Evangelista. It's available from *Backwoods Home Magazine*, 1-800-835-2418.)

Alternative lighting

Decide on what emergency lighting you will use. A lot of folks prefer to buy a generator to power their entire home during a short to moderate power outage. In our remote setting, we prefer not to have to depend on generators running 24 hours a day; we use one for an hour or so and then only as truly needed. Remember, gasoline or diesel fuel may be exceedingly hard to get. Don't waste it for luxury.

We have used propane wall lamps, which give good light and are cheap to run; Coleman lamps, which also give good light, but must be pumped up to pressure from time to time; and kerosene lamps. Candles can do once in a while, in a pinch, but are easily upset, and are dangerous to use.

Be sure you have a good supply of matches, as you'll need more than you think. A few butane lighters also come in handy. I also keep a flint, steel, and magnesium fire starter, just in case. The magnesium burns very hot so it will light a fire, whether your fuel is wet or dry.

Drinking water

Have enough food and emergency water stored up if you don't have an alternative water supply, such as a well with hand pump, spring, or cistern. **Never** drink ***any*** surface water, such as from a lake or stream, that is not either filtered with a good filter to remove contaminants such as giardia, which causes terrific diarrhea, or boiled for five minutes after straining off the major silt. You can treat questionable water with iodine, in drop or tablet form, but personally, I think the taste is awful. I prefer boiling, cooling, then shaking the water container vigorously to re-oxygenate it, which dramatically improves the taste. **Never** drink any water from a lake or stream in or downstream from agricultural land (leaching nitrates from fertilizer) or factories/mines (heavy metals, toxic wastes), as these may escape all but the most expensive filters. Boiling will not make the water drinkable.

You can obtain water for washing and flushing the toilet from several sources: melting snow, ditches, ponds, livestock tanks, rivers, and other surface water. These will all do the job. When your water goes into an urban sewer, be sure you can "dump flush" the toilet, however, as some systems depend on pumps (electricity) to move the sewage. Check it out *before* you need to. (See chapter 7 on storing water for an emergency.)

Vehicle emergencies

A lot of our day-to-day emergencies involve being stranded with our vehicle. A mechanical failure, a flat tire, being stuck in the snow, stuck in the mud and so on can make life miserable. We need to make it less miserable—and not life-threatening as it can sometimes be.

Vehicular emergencies are compounded by other emergencies. If you are stranded in a blizzard, possibly because of an evacuation or civil disturbance, it can quickly lead to a life-threatening situation. You should prepare for these situations also.

It may seem basic, but always know where you are, and know the best route to take to get to where you are going. Carry and use a map.

In cities, avoid "bad" neighborhoods, even if it requires going out of your way.

Pay attention to weather conditions when you drive, and prepare accordingly. It's shocking to us to see folks in cars during cold, unpredictable win-

Vehicle kit

- Jack & lug wrench
- Spare tire
- Shovel
- Battery jumper cables
- Basic tool kit
- Fix-A-Flat
- Oil
- Lighter air pump
- Gallon of drinking water
- Blankets
- Basic first aid kit
- Flashlight
- Emergency food
- Candles with matches
- Map
- Cell phone or C.B. can be a life saver.

ter weather, dressed in shorts and sandals. That's playing Russian roulette. Wear warm clothes during cold weather, or at least have them with you, including warm socks and boots.

Even in summer take some warm clothes along with you. Some evenings get downright cold in many areas, especially when there is a drizzly rain pouring all night.

If the weather is really bad, don't drive. Seems logical, but a lot of unprepared folks set out into the teeth of a blizzard or hurricane because *nothing* is going to disturb their plans. Waiting a day or two is a lot safer. Listen to a weather radio or current weather forecast before setting out. Even a four wheel drive can run into icy roads or drifted snow and have trouble.

If you do get stuck or stranded by a mechanical failure in bad weather, stay with the vehicle. A lot of folks lose their lives by trying to hike for help. In a vehicle, there is absence of wind, making comfortable survival much more certain. Your candle can heat the inside, warn other motorists of your presence, and attract the attention of rescuers. **But be careful!** Everything inside a vehicle ignites easily and burns with toxic smoke. Keep a down-wind window cracked to prevent carbon monoxide poisoning, especially if you run the vehicle (with the tailpipe in the clear) from time to time, to keep warm. (See Chapter 10 on emergency gear for your vehicle.)

A plan, a truck, and a survival pack for short-term evacuation

There are many situations where you may be suddenly called upon to evacuate your home: forest fire, hurricane, earthquake, chemical spills by truck or train, flood, or even radical emergencies such as terrorist activity or riots. To prevent panic, be prepared before anything of this sort happens.

We also have our favorite photo albums, important papers, and keepsakes localized for instant grabbing.

Have a "plan of action" worked out, discussed with all family members. Where you will meet? Who will carry what, and where you will go—a relative's house, a campsite on state or federal land?

If you always keep on top of the weather, the news, and local conditions, you will have a jump-start on evacuation, allowing you to prepare beforehand and avoid panic.

Have the car/truck fully gassed up and carry some reserve gas in cans....just in case. Carry not only your "grab and get" supplies, but be sure you have maps and your vehicle emergency equipment, as well. Try to stay out of main traffic areas/jams and stay out of dangerous areas, i.e., near the path of a fire, the chemical attack, the riot, etc.

It's a good idea if every member of the family has his/her own survival pack. Even a cheap backpack containing warm clothing, food, water, a lighter, candle, first aid supplies, etc. will provide extra security if family members become separated or if someone forgets something.

A survival retreat for long-term evacuation

All families today should have a safe place they can evacuate to should things get dangerous in their home area. This is especially true for city-dwelling people. Such possible realistic scenarios as riots and severe economic depression or collapse, which, considering the current global economic mess, may be the most likely scenario of all.

This retreat can be called a summer home, a vacation home, hunting cabin, or whatever. Calling it a survival retreat immediately brands you as a nut-case radical. Just quietly go about your business, not making waves or attracting attention, and secure your hideaway.

This safe spot should be located at least an hour's drive from a city—two hours from a large urban area such as an L.A. or Chicago-type metropolis. A rural location is great, but a more remote place is even better. Some folks freak out at this, preferring a small rural town setting. That'll work, too, as a close small community can work together. But we've found that that is not always the case, and we prefer to keep to the boonies.

You want something with a little acreage in case you must raise all of your food. Some people claim you can raise all your family's food on a small 50 x 50-foot garden. Nope, you cannot! When you must can all winter's food, raise all of your small grain and corn, and grow feed for any livestock you have (chickens, rabbits, etc)—and take into consideration weeds, drought, and predation—you will need more land, at least two acres to feed your family well.

Your retreat needs to have non-electric capabilities (or alternative power such as solar, wind, hydro, or generator—see Chapter 8 on electricity), well or spring water, wood heat, and alternative lighting, with appropriate fuel stored up in advance.

The sooner you get started the better, for you can plant an orchard, berries, and perennial vegetables, such as Jerusalem artichokes, asparagus, etc., allowing time for them to mature to bearing age. You can also have adequate time to slowly stock up on food and other supplies, work the garden plot up very well, fence any land you need to, develop irrigation, etc.

Where should the survival retreat be located? That's largely a personal preference. We love the very remote country: central Alaska, northern British Columbia, the Montana and Wyoming and Idaho high country. But it's certainly not for everyone. Many more folks would be happier with the plains states: Maine, Arkansas, Missouri, or Minnesota. You can still find very rural areas, far from a freeway or big city, and find it easier to raise food for the family.

And the survival retreat does *not* need to be expensive. I have personally priced many retreats, in diverse locations, nationwide, priced below $35,000.

Choose the retreat with care, taking into consideration such things as nearness to a freeway (dangerous refugee traffic in some scenarios), nearness to a military or other strategic government location (possible missile or terrorist strike), nuclear plant (possible malfunction or terrorist strike), below a dam (possible rupture and flooding), or near any factory, chemical plant, etc.

The decision to evacuate

The decision to evacuate for long term, perhaps forever, of one's home is stressful and highly personal.

It's best to leave before a situation becomes very serious. There are several reasons for this. First, there will still probably be gasoline available. The roads will be relatively uncongested, as most people have a tendency to wait until the last moment to leave, hoping desperately that things will get better. Travelling will be less stressful and much safer. A hoard of panic stricken, angry, hopeless people, even the "nicest" people in the city, could become dangerous.

Take a good map and trace out a route, avoiding freeway-congested areas and large cities, even if it means going miles out of the way. We would prefer to stick to backroads and minor highways. Drive this route often during normal times to familiarize yourself with it and work out any possible problems.

Your goal today should be to learn to be more self-reliant and less dependent on large societal systems that could break down in a major catastrophe. I recommend you subscribe to *Backwoods Home*, PO Box 712, Gold Beach, OR 97444, 1-800-835-2418, $21.95/year. Think of it as enrolling in a college level course in preparedness. Δ

EVACUATION NEEDS

Storage food in large cooler #1

- Instant potatoes
- Dry milk
- Canned tuna
- Dehydrated eggs
- Dry noodles
- Flour
- Shortening
- TVPs
- Dry soup mixes
- MREs (military instant Rice meals; meals ready to eat)
- Dry beans
- Margarine powder
- Dehydrated fruit
- Dehydrated vegetables
- Tomato powder
- Baking powder
- Salt
- Spices & condiments
- Pudding mixes
- Cornmeal
- Instant coffee, tea, drink mixes
- Sugar

Kitchen box in large cooler #2

- Frying pan
- Large pot
- Smaller pot
- Mixing bowl, steel (can double as cooking utensil)
- Matches & lighters
- Toilet paper
- Paper towels
- Dish towel
- Dish soap
- Candles
- Dish scrubber pad
- Bowls for family
- Silverware for family
- Metal spatula
- Roll of duct tape
- Small roll of wire
- Metal cups for family
- Small water filter
- Propane stove & tanks
- Flashlight & batteries
- Hatchet

Medical Kit (as previously detailed)

Sleeping Gear (in large plastic box)

- Sleeping bags
- Candles & lighters
- Coleman lantern
- Unopened gallon of lantern fuel
- Bow saw
- Warm socks & jackets
- 10′ x 12′ plastic tarp
- lightweight tent
- Radio

Rifle/shotgun and ammunition (food procurement, signaling, and family protection)

Personal backpacks

- Warm clothes
- Emergency food
- Socks
- Stocking hat
- Basic fishing gear without rod
- Small first aid kit
- Space blanket
- Flashlight
- Roll of wire & rope
- Pocket knife
- Canteen with cup
- Lighter
- A few dollars in quarters & bills

Chapter 2
Storing LP gas, gasoline, diesel, and kerosene

By Emory Warner

Home storage of fuel is a good idea, but it must be done properly. Your car, truck, tractor, and standby generator require fuel. So do your propane refrigerator and kerosene lanterns. Here are appropriate methods of storage for LP gas, gasoline, diesel fuel, and kerosene, as well as some tips on safe fuel handling.

Fuel types

LP gas is one of the easiest fuels to store and also one of the most dangerous. It is a highly versatile fuel which can be used to power internal combustion stationary engines, tractors, and other motor vehicles, as well as for cooking and heating. LP has two serious drawbacks: First, it must be stored under pressure to remain a liquid; any leak (which may not be visible) could leak away all of your fuel without your knowledge. Second, LP is only slightly heavier than air, and will disperse at the exact ratio to produce an explosion. It will also "puddle" in low spots, waiting for an ignition source.

Gasoline has the advantage of being a liquid at room temperature. But it is probably the hardest fuel to store for any length of time. It has a high vapor pressure (which means it evaporates quickly) and will go stale in a few weeks if not chemically treated. It does have a fairly high ignition temperature (about 1100° F) even though it does not need a large volume of heat to ignite. Stored gasoline must be treated with a BHT additive like *Sta-Bil* and protected from moisture if it is to be stored for any length of time.

Large quantities of gasoline make me nervous. I used to live on the water in southern Maryland, and was witness to several boat explosions and fires due to gasoline vapor in the bilges.

Kerosene is one of the easiest fuels to store, and is more versatile than most people think. It does not evaporate as readily as gasoline and will remain stable in storage with no special treatment. Many pre-1950 farm tractor engines were designed to run on kerosene, and diesels will run on kerosene if necessary. Kerosene stoves and refrigerators are also available and would definitely be preferable to LP models from the safety standpoint.

Diesel fuel stores almost as easily as kerosene and is becoming more and more popular among the self sufficient. It is difficult to ignite intentionally and almost impossible to ignite by accident. Two grades are available: #1 diesel which is old-fashioned yellow kerosene, and #2 diesel which is the same thing as #2 home heating oil. (You may see literature to the contrary, but #2 diesel is #2 heating oil. Period.) Diesel fuel presents its own unique storage problems: The first is that it is somewhat hygroscopic; that is, it will absorb moisture from the air. The second and related problem is sludge formation. Sludge is the result of anaerobic bacteria living in the trapped water and eating the sulfur in the fuel. Left untreated, the sludge will grow until it fills the entire tank, ruining the fuel. Stored diesel fuel should be treated with a biocide like methanol or diesel *Sta-Bil* as soon as it is delivered. Unique to #2 is the fact that some paraffin wax is dissolved in the fuel and will settle out at about 20° F, clogging the fuel filter. This "fuel freezing" may be eliminated by adding 10% gasoline or 20% kerosene to the diesel fuel. Commercial diesel fuel supplements are also available to solve the same problem. Diesel should be filtered before use.

Alcohol (ethanol) is not commonly considered a storage fuel, but here is the data on it for those who distill their own. Alcohol is as hygroscopic as it gets, and must be stored in a sealed container to prevent moisture contamination. It is about as volatile as kerosene and presents the unique problem, when ignited, of burning with an almost invisible blue flame. It may be best to store the raw material for stilling the alcohol and producing the fuel as needed, rather than producing a large quantity and storing it.

Whatever fuel you store, it would be a good idea to monitor your fuel usage and plan your storage around a 90-day supply.

Safe fuel handling

Regardless of the fuel in question, all liquid fuels should be handled in the same matter as the most volatile, which is either gasoline or LP gas. Fuel should be stored in an isolated area, downhill and downwind from any other buildings. Fuel vapors are heavier than air, and will flow downhill. LP tanks should be left in the open and not enclosed in any way. Liquid fuel tanks can and should be stored in a well-ventilated building or open lean-to to prevent solar heating from evaporating the fuel. If the storage location is permanent, consider using a buried tank. If set below the frost line, temperatures are stable at 55° F or so, which will inhibit evaporation. The tanks will be safe from everything, including stray (or aimed!) gunfire, brushfires, and just about everything else except the EPA. If buried fuel tanks offend your sense of environmental responsibility, then consider an underground vault. This has the added advantage of being able to inspect the tanks from time to time.

Regardless of the tank location, a dry chemical or CO_2 fire extinguisher should be hung on the outside of the building or near the pump. Any electrical fixtures should be "explosion proof" (sealed) and wired in sealed conduit to prevent fuel vapors from coming into contact with electrical sparks. Prohibit smoking or carrying of smoking materials within 50 feet of the fuel pumps. Electrical fuel pumps should have a heat sensitive shutoff to stop the pump in the event of fire. Always shut down the engine of the machine being fueled. Promptly clean up any spills. Last of all, be certain to use only the equipment that is approved for the fuel in question. (Some fuel pumps are approved for diesel only, and are unsafe to use for gasoline.)

Fuel storage methods

Liquid fuels use the same storage systems and will be covered as a group. LP gas is normally stored in pressurized tanks supplied by the LP dealer, and will be only briefly covered.

The most basic fuel storage system is the common portable fuel can. If you are still on the grid and have a job "off the property," then this is a workable and economical method of fuel storage. A minimum of three cans will be required: one full at all times, one for use as needed, and one to be refilled at the first opportunity. Rotation of the cans will ensure some amount of reasonably fresh fuel at all times. This storage system has the added advantage of portability in the event that the storage site must be abandoned. Use only approved containers, and use caution not to mix up containers. The standard color code for portable cans is blue for kerosene, red for gasoline, and yellow for diesel fuel. This is not cast in stone. Use whatever color scheme you like, but be consistent with it. Gasoline introduced into a diesel tank will make the diesel engine hard to start when hot. Gasoline in a kerosene heater will explode like a Molotov cocktail. Diesel #2 in a kerosene lamp will smoke and stink and soot up the globe. If you use all three fuels like we do, it seems that you will be filling a fuel can every time that you go out. Delivered fuel is much more convenient, and usually cheaper.

The next storage system is the 55-gallon drum used with a hand pump or horizontally on a rack. This is a highly flexible storage system, as drums may be added as needed to suit individual requirements. Most fuel dealers have a 100-gallon minimum delivery, so at least two drums will be needed. You can even load one drum in your truck, drive to the service station and fill it, then bring it home and pump the fuel into your storage drum. Drums are also portable enough in the event that the storage site must be abandoned. The only disadvantages are the negligible cost of the drums and that the drums will eventually rust and leak. We use drums for our kerosene and gasoline storage. Label each drum clearly if you are storing more than one type of fuel.

If you wish to store large quantities of fuel, then the built-for-the-purpose fuel tank is the system of choice. Tanks are available new in capacities from 100 to 10,000 gallons in above ground and underground types. Most commonly used here in the Northeast is the standard residential 275-gallon fuel tank. These are available new at plumbing and heating suppliers for about $150. Used tanks are usually available free for the hauling, including whatever fuel is in them. As a side note, an individual with a pickup truck and a reciprocating saw could make a fairly decent living removing old fuel tanks as homeowners change away from fuel oil to natural gas. This is about the dirtiest work available, and pays about $100 per tank. The removed tanks could be cleaned up, painted, and resold for $50 or more. I have accumulated about five or six tanks in the last few years without really looking for them.

Fuel dispensing is a matter of choice. An elevated tank needs only a valve and filter; gravity will do the rest. We prefer to use hand pumps for our kerosene and diesel tanks. Valves have been known to leak, and vandalism is an unfortunate reality of modern life-especially if the vandal elects to open the valve on a tank of gasoline and follow it up with a lit match. Hand pumps are safer, and they are more easily secured if the tank must be left unattended.

The author's system

My personal fuel storage system is a salvaged 275-gallon fuel tank with a hand pump and filter for our diesel fuel storage. Our principal tractor is diesel powered. We also use it to operate a PTO (power takeoff) generator for standby use. We use two or three 55-gallon drums for kerosene storage, with a lift pump for dispensing. (We rely on kerosene heaters to supplement our woodstove.) But, as I have a job "off the property," and we have two old gasoline engine tractors, as well as a chainsaw, lawnmower, etc., the fuel can system works well for our gasoline supply. This is particularly suitable for us, as I feel uncomfortable about storing large quantities of gasoline. Δ

Chapter 3

Canning meats and vegetables at home—it's not only easy, it's safe and inexpensive

By Jackie Clay

While quite a few people still put up pickles, jams, jellies, and tomatoes, it is estimated that less than 5% of the population in the United States actively cans vegetables, meat, fish, and poultry. Why? I think it's because people fear it is difficult, dangerous, and expensive. They're afraid they will give their families food poisoning, and they think they can buy canned goods cheaper at the store.

Let's look at the arguments realistically. Difficult? I can put up 10 pounds of meat in less than 2 hours, while I work on the word processor or home school our son, and I'm no rocket scientist.

Dangerous? No, the canner won't blow up if you read common sense directions and regularly monitor the pressure, adjusting the heat as needed to keep it at the correct pressure. Nor do you have to worry about tainted food if you follow the precautions given in a canning book.

Expensive? If it was, this frugal home canner sure wouldn't do it. On an average, it costs me 10¢ to can a jar of vegetables and meat (provided I grow the produce and hunt or home raise the meat) and even less if I can on our wood range, which I often do during the cool mornings of autumn.

And that jar can be anywhere from a half-pint to a half-gallon of food. Store-bought canned vegetables and meat cost a lot more than that. Just yesterday I priced eight ounces of canned chicken breast, on sale, at $2.19. That translates to $4.38 a pint vs. 10¢ for a pint of my home-canned chicken. And even if I bought the chicken from a butcher, then canned it, I could put up a pint for half the cost of that store-bought canned meat. And besides the reduced cost, there are no chemical additives in anything I can.

To successfully can, all you need are some basic equipment and instructions and you can enjoy clean, chemical-free, inexpensive, and nutritious vegetables and meat all year long.

Equipment

A good **canning book** is a must for all home canners. I have four, not because canning is difficult or that I am stupid, but because each provides a lot of different recipes and ideas. The processing and safety tips in each book are the same, but I'm always open to new ideas and you should be too. But until you become experienced you should not free-lance, that is, change the recipes, as incomplete processing times can result.

Jars are a must, of course. Some folks will swear that you cannot process meat and vegetables in other than brand-name canning jars and that if you use pickle, mayonnaise, salad dressing, or other jars, which a canning lid and ring will fit well, they will break in the pressure canner. Not so. I've used these "orphan" jars for over 35 years, along with Mason and Kerr jars. I can see absolutely no difference in the instance of breakage. So a note on a bulletin board or in a free shopper, or just plain telling everyone you know, will usually result in a lot of free jars.

Examine each jar as you wash it at home for any minute chips or cracks, especially in the rim of the glass. Chips will usually result in the jar not sealing or a larger crack developing. And cracks, of course, weaken the jar and cause it to break as it processes.

A **pressure canner** is necessary to can all meat, fish, poultry, and meat products such as soups, stews, spaghetti sauces, etc. This is a fairly large expense for many frugal folks, costing about $125 for the larger, more work-worthy size. But, when you figure it will last for over 20 years, without maintenance, it is one of the best buys of a lifetime. Remember, you can use it to put up nearly anything that you would see on a store shelf or that you may hunt or fish for yourself.

Vegetables and meat products must be processed in a pressure canner to raise the temperature of the product you are processing and hold it at that level for for a considerable time. This ensures that you will kill all the bacteria.

Water not under pressure, as is used in simple water-bath processing, boils at 212° F. This is fine when canning fruits and tomatoes which have high acid content that kills microbes that may survive the boiling. But it is not adequate for low-acid vegetables and meats. Still, it was the method used by our grandmothers, as they did not have pressure canners as young women. And the food they canned did seal and was usually okay to eat. Usually. But I won't gamble my family's lives on "usually." Food poisoning is nothing to fool with, so I, and all intelligent home canners, process all meat and vegetable products in a pressure canner.

Jar rings (sometimes called bands) and lids are a basic, as well. Jar rings are used to hold the lid in place during processing. They *do not* help keep the jar sealed during storage. A properly sealed jar will remain sealed, without its ring, even when handled. In fact, jars should *not* be stored with rings on them as dampness can collect under the rings and promote rusting, making the ring useless for further use, and the jar lid may rust too, which will ultimately cause the seal to fail and the food to spoil.

Jar lids need to be of high quality. Never use el-cheepo lids from Asia that you've never heard of before. The three most dependable brands are Mason, Kerr, and Bernardin. The lids are boxed a dozen to a box, and they consist of a disk of lightweight metal, rimmed with a rubberized compound which, under heat, effectively seals the jar. They are *not* reusable and should be discarded after one use. Boxes of lids will store for years, remaining good. Self-reliant people should stock up on jar lids.

Other handy equipment to have around are a **canning funnel, sharp knives, mixing bowls, a jar lifter, chopping board, and a lid lifter** which neatly picks individual jar lids out of a pan of boiling water.

Canning steps for vegetables and meat

1. Have all the equipment on hand and ready.
2. Inspect the jar rims again for nicks.
3. Fill the jars.
4. Wipe off the jar rim.
5. Put the lid into place.
6. Screw the ring on firmly, but not forcefully.
7. Place the jar into canner.
8. Put the canner lid on, securing it firmly, but leaving the exhaust vent open.
9. When a steady stream of forceful steam comes from the vent, close it off.
10. Wait for the pressure to build to correct readings, then begin counting the processing time.
11. When time is up, shut off the heat, then allow pressure to drop to zero.
12. Remove the canner's lid away from yourself, so steam does not scald you, and remove the jars.
13. Set jars on dry, folded towels away from drafts until they cool.
14. Examine for seal using one finger to press on center of the lid. If it gives, it is not sealed and you must reprocess it using a new lid.
15. Remove the rings, wash the jars, and store them in a cool, dark, dry place.

While many foods are most easily canned using the hot pack method (where partially or wholly cooked food is placed in hot jars, then pressure canned), most foods I can are placed in the jars cold for ease and speed of processing a batch. Read your canning book, then decide which method is best for you and the food you are processing.

When getting ready to can a batch of food, have all your equipment clean and ready to go. The jars do not need to be sterile but must be freshly washed and clean. It is good to remember that in canning, cold should not be mixed with hot. That is, don't put boiling food into cool jars, cold food into hot jars, or set hot jars on a cool surface. I learned a lesson after many years of canning: every once in a while, a jar bottom would break during processing. Finally, I discovered that if I warmed up the canner before setting warm jars of food into it to process, I drastically reduced this breakage. Just turning the burner on a few seconds before placing the first jars in did the trick. Match the canner's bottom temperature with the jar temperatures.

Place a small pan of water on to boil. Separate the jar lids and drop into the water. Boil the lids, then remove them from the heat, but keep them warm.

Place the jars to be filled on a folded towel, then carefully fill each jar. The folded towel not only moderates the temperature from the table or counter surface, but it also catches spills making cleanup a snap.

Using a canning funnel helps keep foods from dripping onto the jar rim. You want to prevent this, especially with meats and poultry, as grease on the jar rim (or even a tiny bit of green bean) will keep the jar lid from sealing onto the jar rim correctly. An unsealed jar equals spoiled food.

Cut-up or whole green beans, potatoes, corn, other vegetables or meat, poultry, and fish may be placed in the jar raw. This is the raw pack which I most often use. Canning books have gotten away from raw-packed meats. I believe it's because the writers felt that home canners would become sloppy and possibly cause incomplete processing, resulting in meat which might harbor harmful bacteria. It is possible, but personally I get tired of folks trying so hard to keep me safe from my own responsibilities. And when I have an elk to can—and an elk is a lot of meat—I need to get it processed fast. So I still raw pack pieces of boneless meat. I am not advising others to do what I do; I am only explaining how I do it. You may well choose to hot pack partially cooked meat.

I place fat-free boneless steaks, roasts, stewing meat, and just plain chunks of meat into a clean jar. (I use everything from half pint jars to half gallon jars, but I always process jars of a like size together—I don't mix sizes.) A teaspoonful of salt may be added but is not necessary. Water is not usually added, so the jar rim is carefully washed with a warm damp cloth, the hot lid is put in place, and the ring screwed down securely but not overly tight. The jar is now ready to put into the canner.

Tips for canning meat and vegetables (Low acid foods)

1. Always use a pressure canner for all meats, fish, poultry, wild game, vegetables, and products containing these products such as soups, stews, sauces, etc.

2. If unsure of processing time, process the jars for the ingredient which requires the longest time. For instance, spaghetti sauce needs to be processed for the time given for meat, not tomatoes.

3. Can only fresh food. Never use questionable food for canning.

4. Remember, hot + cold = broken jars.

5. Never take shortcuts in processing time.

6. Following canning book directions results in wholesome, long-keeping canned food.

7. Get into the habit of checking and rechecking for nicks and cracks in jars. You'll save frustration, food, and money.

8. Don't try to pressure-can with arty jars that use zinc lids, or glass tops with wire bails, etc. You can't tell if they are sealed or not. A dangerous practice.

9. Don't experiment with recipes for canning until you are experienced and understand canning fundamentals completely. There are hundreds of tried and true recipes out there for you to gain experience with. (Spices may be varied to one's taste without endangering processing.)

10. Don't can meat with bones or fat intact, excepting fish or poultry. The bones and fat impart an unpleasant flavor at times, especially in game meat or mutton, and the bones take up unnecessary room in the jars. Also remember that fat is the #1 enemy of jars sealing.

11. Before eating or tasting a newly opened jar of food, visually check it for normal appearance and odor first. If there is any frothing, cloudy juices, unusual odor, or if a jar gushes or is not sealed when opening, then discard the contents where animals cannot get at it. This food is a definite risk. When it passes inspection, boil it for 15 minutes, just to be sure. This will kill bacteria that might make you sick.

Hot-packed meat, such as partially cooked roast, steak, stew meat, boiled chicken and meat products, such as stew, chili, soup, etc. are put into warm jars. Liquid is usually added, i.e., broth or soup, the rim carefully wiped, the hot lid placed on, and the ring tightened.

Hot packing is great and convenient for canning large batches of spaghetti sauce, chili, stew, baked beans, canned dry pinto beans, etc. Just cook and dump into jars, then process. Okay, I'm simplifying, but once you get the hang of it you'll see it becomes that easy.

All raw meat should be heated or "exhausted" in the jars, which are placed in a pan of water deep enough to heat the jars thoroughly, while the water boils, but not so deep that the water boils into the open jars. Bring this pan to a slow boil and check with a meat thermometer inserted into the center of a jar. You need to heat the meat to 170° F, then quickly remove the jars from the bath with a jar lifter, place them on a folded towel, wipe the rims clean, and put the lids and rings firmly into place. Then place the jars in the canner and exhaust the canner. This means you should ensure there is a steady stream of forceful steam escaping the vent, not just spurts now and then.

After the canner is hot, i.e., exhausted, close the vent and begin raising the pressure until it reaches the desired processing pressure. Remember that most canning books give an average processing pressure of say 10 pounds. But if you live at an elevation higher than 1000 feet, you must bring the pressure up higher. Check your canning book for your exact pressure needs. Begin to count the processing time.

Keep the pressure at the correct reading by adjusting the heat under the canner or moving the canner gently on a wood range's surface or adding wood to the fire, as needed. If you let the pressure fluctuate, it will suck the fluid out of the jars. The resulting food will still be edible, but may be dry-tasting, or food bits may get under the jar lid making a proper seal impossible.

After the food has processed long enough, turn off the heat or remove the canner from heat. Allow the pressure to return to zero, then carefully remove the canner's lid—away from you, so escaping steam does not scald your arms or face. (Don't get in a hurry, thinking to just leave the jars in the canner with the lids on to cool. The jars will not seal correctly.)

Carefully set the hot jars, still boiling and bubbling, on a dry (never damp or the jars may crack) folded towel in a draft-free area to cool. Soon the telltale musical "pings" will let you know they are sealing. Never fool around with the hot jars or you may disturb the seal.

When the jars are perfectly cool to the touch, remove the rings, and wash them for next time. Then wash the jars with warm soapy water, rinse and dry them, and then store them in a cool, dry, and dark place.

Canning green beans

Pick the beans, wash them in cool water, and prepare to can them immediately. The beans may be canned whole, Frenched, or however your family likes them. I usually can a variety, from whole to Frenched, with the bulk cut into convenient chunks an inch or so long. Cut the beans, removing any tough strings, as well as the stem and pointy end, if desired. Using a canning funnel, dump the raw, cut beans into clean jars placed on a folded towel.

Pour two inches of water into the canner and place the basket or inner kettle into place. The canner must never boil dry or it will warp.

In the meantime, have enough lids separated and boiled. Also, have boiling a large pan of water with which you will cover the beans.

Fill all jars to within one inch of the rim. This is called "head space" and is necessary for proper processing and storage. In canning, you do not want to cram as much food into a jar as it will hold. Some foods expand as they process, and all need a certain amount of head room to process and keep well. Always follow your canning book's directions exactly.

You may add a teaspoon full of salt to each jar if you want to enhance the flavor, but it is not necessary.

Pour boiling water into each jar, just covering the beans. Then carefully clean off the rim of each jar with a warm, damp cloth to remove any food bits which might prevent the jar from sealing, and check for nicks in the rim with a clean finger. Place the lids and rings into position. Do not over-tighten the rings. The ring only holds the lid securely into place for processing, and does not have anything to do with how well the jar seals.

Bring the canner up to the same approximate temperature as the jars, then carefully place the hot jars into the canner, taking care not to thunk them together. Leave space between jars to allow for steam to circulate during processing.

Tighten the canner lid, raise the heat to high, and allow the canner to exhaust. When a steady, forceful stream of steam blows from the vent, close it and let the pressure build up. When it reaches the correct pressure (10 pounds for altitudes less then 1,000 feet above sea level, but see your canning book for higher elevations), begin timing. Pints of green beans will be processed for 20 minutes, and quarts for 25 minutes.

At the end of this time, turn off the heat or remove the canner from the heat and allow the pressure to return to zero. When it does, unfasten the lid and carefully lift it away from you, allowing hot steam to escape away from your arms and face. Then remove the jars carefully with jar lifter, again not thunking them together, which could result in cracks. Place the hot jars on a dry folded towel in a draft-free area until they cool.

Canning ground meat

Any ground meat or ground meat products such as chili, spaghetti sauce, taco meat, etc., should be cooked before it is canned or it will not have a good texture. The meat will clump together in lumps. So, in a large frying pan fry the meat in as little grease as possible. (Grease is the #1 enemy of jars sealing). Add the spices you desire, then the tomato sauce, beans, chopped onions, or whatever.

Some reliable food processing books

The Ball Blue Book (the Guide to Home Canning and Freezing), Alltrista Corporation Direct Marketing Department PK31, P.O. Box 2005, Muncie, IN 47307-0005, $5.95, including shipping. IN rcsidents add 5% tax

Putting Food By, by Hertzberg, Vaughan & Green, Stephen Green Press

Stocking Up, by Carot Hupping, Rodale Press

Other great books are available, of course. Check out your local book store (or their catalog) and the library. Your county extension office, usually located in the courthouse, can usually provide free (or very low cost) canning publications and leaflets.

Have clean canning jars on hand, kept hot in water. Also have a sufficient number of boiled lids on hand so that the process proceeds as quickly as possible.

Using a canning funnel, carefully fill each jar to within an inch of the rim (one inch head space), wipe the jar's rim with a warm, damp cloth and inspect the jar again for any minute nicks.

Then screw the band snugly on over the hot lids and place each hot jar into the warm canner. Again, be careful not to thunk the jars together as it could crack them.

Turn up the heat with the canner vent open, and wait until a steady stream of forceful steam exits the vent. When this happens, close the vent and wait until the pressure arrives at 10 pounds. (Again, if you are canning at altitudes over 1,000 feet above sea level, check your canning book, as the pressure must be increased with the increased altitude.) When the correct pressure is attained, begin to count the processing time. Be careful not to let the pressure fluctuate as it can blow liquid out of the jar.

Pints will be finished in one hour and fifteen minutes and quarts in an hour and a half. I process half pints, which are very handy for casseroles, etc., for one hour and fifteen minutes.

When the jars have processed for the correct amount of time, turn off the heat or carefully remove the canner from the heat, and allow the pressure to return to zero. (Do not try to hurry this by fooling with the exhaust valve or you may end up with broken jars or jars that do not seal.)

Carefully remove the cover away from you to avoid steam burns, then take the jars out carefully and place them on a folded, dry towel in a draft-free place to cool.

When completely cool—overnight is best—remove the rings. Then carefully wash each jar in warm soapy water, dry them, and store them in a cool, dark, dry place. This meat will last indefinitely, regardless of what you have been led to believe. Δ

Chapter 4
Long-term food storage

By Jackie Clay

To avoid panic and discomfort, it is provident and wise to stock up on food for not only survival, but reasonable comfort and happiness, should we need to live off what we have stored in our pantry, root cellar, basement, or attic.

We should also rotate the foods we store in order to have wholesome foods to choose from. But just how long are foods actually good?

Some items at the store have a "freshness date" and it is commonly believed that after that date the products will not be good. And even preparedness companies cite a shelf life of five years in their storable foods. Then along comes some strange person, such as myself, who tells a different tale. As a long-time survivalist and home canner, with nothing to lose or gain from telling you anything but the truth, you might listen to my experiences.

I have always kept at least a two year supply of food stored against bad times, whether it be an illness, injury, loss of a job, storm, or worse. This is a practice I learned from my parents and grandparents who lived through and learned from the Depression. Every year I home-can hundreds of jars of food, most filled with home-raised produce and meat, some with meat from hunting, some with items purchased at great sales at the market throughout the year.

In one year we canned two deer, a tremendous tomato crop in another, a bumper apple crop in yet another, and so on. I always can all I am able, as in other years the crop may not be so good and the hunting may be sparse. In this way, my pantry leapfrogs, as we do not consume all of last year's canned food. So, through the years, the canned goods build and build, and despite rotating the shelves to try to use up the oldest, our supply expands.

Likewise, other pantry supplies, bought from the stores, grows and grows as one great sale follows another.

Okay, the bottom line: Just how long will this stuff keep? Do I really have to throw it to the chickens after a year? Two years? Five years? The answer is one word.

No.

Canned goods

No matter what you read in canning books (the newer ones, of course), on labels, in magazines, and no matter what your neighbor or friend tells you, canned foods will last nearly indefinitely.

Now, you *must* store all canned foods, including home canned foods, in a cool, dark, dry place for optimum shelf life. Storing them in hot, light conditions will sometimes result in changes in texture, color, and taste as well as hasten the breakdown of vitamins. (It is this breakdown in vitamins that most often gives the warning, which sounds so dire: *use before such and such a date.*)

It is true that most canned foods will lose some vitamin content. But if you've ever been hungry—I mean real hungry—you don't worry if the vitamin C in the canned tomatoes is below national standards. Besides, we figure we make up any vitamin shortfall with the fresh produce we eat nearly every day from the garden.

Storing canned foods in damp conditions, as often found in basements or root cellars, can shorten the shelf life, and sooner or later the cans and jar tops will rust, weaken, and the contents will spoil. If this is your only storage facility, be certain to use up any cans or tins that are beginning to rust before they go bad and always check such containers for mold, cloudiness, odor, or an unsealed or bulging condition. All indicate spoiled foods. Likewise, boil all vegetables or meats for 15 minutes to kill pathogens, even if not apparent. Just to be sure.

I have home-canned jars of food that are at least 20 years old, which we use from time to time. For instance the cherries we picked from Dad's orchard, which we parcel out frugally until we get our own trees bearing. These foods taste, smell, and look great, despite their age. Plenty good for an emergency situation, for sure.

Dry goods

Okay, let's move on to the more nebulous items, such as dry goods, like flours, dry milk, sugar, etc. Will all of these store indefinitely as well? Yes and no, depending on the product. Let's start with those that have an extremely long shelf life, given good storage practices. By this I mean kept dry, sealed, and stored in a fairly cool, dry, dark location.

Beans, dry peas, wheat, and other dry grains, unprocessed, will keep in storage a long, long time. I have some beans that are more than 700 years old, and they still germinate and grow.

You know I could eat them, if I wanted to. But, of course, I don't as they are treasures from the past.

Because these grains store so long, it is best to store whole grains, including corn, and grind them as needed. For once they are ground, the shelf life decreases, often dramatically. Take whole wheat flour and corn meal for instance. Both of these products can become rancid after a period of from two weeks to a few years, because of the oils in them.

White flour from the store has been "processed," which removes the oily germ and, of course, much of the nutrition. Therefore, it will store for a much longer time than will whole wheat flour. My grandmother did not like to use fresh white flour, preferring to use older flour as it baked better. Right now, I'm using a bin of six-year-old white flour, and it is fine. I do sift it twice to fluff it up because when it sits in the bag for a long time, it settles and packs together. Without the extra sifting, it bakes pretty solid biscuits and bread. Corn meal will usually last, unrancid, for about a year or two in a sealed glass jar.

Other than dampness, a bag of flour or grain's worst enemy is the meal moth. This little bugger is a small, nondescript greyish moth who gets into our grain and lays eggs which hatch out into flour weevils, ruining the flour in a short time. The first sign of weevils are tiny dark specks in the flour, followed by webbing in the can or jar. The moths initially come into our pantry in a bag of flour with a small tear, hole, or unglued section of bag.

Always thoroughly check all new bags of flour or meal at the store, rejecting any that have a tiny leak. Taping the hole at the store is not a cure. Buy solid bags, and immediately get them into good, airtight storage. For long term storage, I put two 25-pound sacks in a good food grade garbage bag, stick a few bay leaves in for good measure, and seal the bag with duct tape. The bay leaves discourage any moths that could possibly get into the sealed plastic bag. These sacks are then either stored in a clean garbage can or sturdy cardboard box, which is also taped shut when full.

I usually freeze five-gallon pails of whole grains in case some minute friends are hitching a ride in our food. The freezing kills them before they become a problem.

It is a very good idea to buy a package of meal moth lures/traps, which attract the moths before they attack your stored flours. The cost is minimal and they do afford good protection. These traps sit discreetly on your pantry shelf, trapping any moths that happen by.

Sugars will last indefinitely. They must be kept dry and sealed to prevent hardening. When I store brown sugar, I dampen a piece of folded washcloth and place it on top of the sugar, then seal the jar. This keeps the sugar from hardening, which is a problem with brown sugar. If a bag or jar of sugar does get hard or crumbly, it is still good, although a bit inconvenient. Just warm up the sugar and add it to the liquid in the recipe to soften it.

Dry milk, dry eggs, dry margarine and butter powder, cheese powder, and powdered cheese sauce are foods that keep very well, if unopened and well sealed. I buy dry eggs, powdered cheese, margarine, orange drink mix, and many other long-storage items from a preparedness company as they are sealed in #10 cans.

I've used some of these foods that were seven-years-old and older and all were perfectly fine. And I've used dry milk from the store which was well sealed and stored for 10 years on our pantry shelf. The milk smelled and tasted normal and resulted in great pancakes, rolls, and sauces.

Home dehydrated vegetables and dehydrated vegetables purchased from preparedness companies in #10 cans make an excellent lightweight, nutritious, long-term storage item. I dehydrate everything from sliced potatoes and corn to tomatoes and peppers. Perfectly dried and securely sealed, they will last for years.

I buy two one-pound foil bags of granulated dry yeast at a time. One I open and pour into a jar, which is stored in the fridge. The other is stored, unopened, in the freezer compartment of the fridge. As yeast only keeps a shelf life of about a year, unrefrigerated, I rotate this yearly, using the frozen yeast to replace the one in the refrigerator at the end of the year and buying a new one for the freezer. But, in an emergency, one can always use a bit of this old yeast or even develop wild yeast to make a sourdough starter.

Salt will keep forever if stored with care.

Baking powder will keep well a long time if stored properly. In fact, the can of Rumford I'm using now was purchased five years ago and it just sits on my shelf. And if it starts to weaken, you can just add a bit more or boost it by adding warm liquid to the mix. Baking soda lasts even longer. I'm using some off the pantry shelf that is nine, count 'em, nine years old. And no one has ever whined about my cooking.

More perishable foods

How about more perishable foods? When we lived on our remote homestead in Montana's high country, we were snowed in for at least six months out of the year, so preparation was a must. We learned that we could stick frozen stick margarine in a cooler we placed in a snowbank and have it last all winter. Unfrozen but refrigerated margarine would keep for about two months, then begin to pick up odors and tastes. We learned that tubs of margarine would keep for nearly all winter in a cold spot on the floor of our pantry, but we did need to protect it from not only our cats and an occasional mouse, but from the dogs as well. Butter lasts a much shorter time, unless kept strictly frozen.

Shortening, bought and kept sealed, will last many years before going rancid. I have used some that was 7 years-old, and it was fine.

Eggs are a big joke with us. Many folks insist on "fresh" eggs, throwing out those a few weeks old. I worked part time for an egg ranch. The fresh eggs were picked up weekly, hauled to a warehouse where they were distributed to wholesale companies, who kept them around awhile before trucking them to super markets where they were finally bought. How much time elapsed? Who knows?

We raise our own chickens but before we snowmobiled our day-old chicks up the mountain one April we had to buy eggs for the winter. We found that if we bought really fresh eggs from a rancher in November, we'd have good eggs in May. I did crack them into a cup, as an occasional one would be bad.

You can waterglass your eggs, but a crock full of those eggs is nasty to reach into. Kind of like dipping into snot for breakfast eggs. It takes your appetite away and it is a bit costly.

We found that keeping the eggs boxed in the fridge or cold corner of the pantry was sufficient to keep them all winter. All eggs to be stored should be carefully inspected for even the most minute cracks as it can allow bacteria to penetrate the egg.

Without a flock of chickens to depend on, it is a good idea to have several #10 cans of powdered eggs on hand to be available in an emergency.

Just a note: home-raised eggs, fresh from the hen, are fine unrefrigerated for many days. I've found hidden nests in the weeds with eggs that have not been sat on by the hen yet, and though they sat out in 90° weather for as much as a week, I used them, finding every one was like it was fresh from the hen.

Meats and meat substitutes

Unless your family is vegetarian, meats in storage is necessary. No, I do not mean in the freezer, as no matter what "bad times" entail, the first thing to go is the power. Lose a job, get injured, not enough to pay the power bill, storms, earthquakes, fires, floods—all can quickly zap the power. While there are steps you can take to keep a freezer from thawing out quickly, they are not enough for a long-lasting emergency.

I have home-canned meat for years, and found it extremely easy, quick, and convenient. Any canning book can help you get started today. This meat, including stews, soups, sauces, fish, poultry, and wild game will keep indefinitely if properly stored in that dark, dry, cool pantry.

Want to store meat before you get that two years' supply of home-canned meat on your shelf? Just look on your supermarket shelves. There's a lot to choose from: tuna, salmon, hash, chicken, ham, sauces, beef, and even bacon. For long-term storage I try to stay away from those convenient "pop tops" with a handy pull ring. They are nice, but can easily get unsealed in the hustle and rush of an emergency. You have to handle and pack them very carefully, or the weakened area that pops can be poked, unsealing them, often without a sign it has happened. Yes, our family does have Spam on the shelves of our pantry, but I handle it very carefully. Soups, stews, canned spaghetti, and so forth, purchased from the store shelves, will also last indefinitely, if kept dry to prevent rusting.

Jerky? Well, to tell the truth, few people ever dry it long enough for safe, dependable long-term storage without canning it as well. In many climates, the meat goes bad or begins to mold in as little as two weeks without refrigeration. If it is dried to a brittle stick, it will keep longer, but it is like chewing on a piece of rawhide. Indians did it, but they were much less fussy than today's urban population.

A popular meat substitute that is lightweight, nutritious, tasty, and long keeping is a product called TVP (textured vegetable protein). You probably best recognize it as the bacon-bits that aren't really bacon. We keep about 15 pounds in factory sealed #10 cans or aluminum bags in our pantry. As most recipes only need about a quarter cup, you can see these lightweight crumbles last a long time.

They come in several flavors: chicken, beef, bacon, ham, and even taco. I've found that keeping several jars of dried soup base next to the TVPs makes a nice couple. Simply adding the flavored soup base to soup, noodles, or whatever, then tossing in the matched TVP, makes a very quick, lightweight, satisfying, and *cheap* dinner, even on the go.

Snacks for storage

Okay, I know goodies may get raised eyebrows, but they sure make an emergency less depressing. Unfortunately, potato chips and other "normal" snacks are primarily grease which turns rancid pretty quickly. But there are still a lot of snacks out there perfect for the pantry. On the top of our list is home-dried fruit. I dry about as much as I can and have gallon jars of dried apples, apple bits, peaches, peach bits, strawberries, pineapple, apricots, pears, and more. I have 10-year-old dried apples in a test jar and I've pulled out a few to nibble on each year for five years now. They are a bit brown, but still very tasty. These dried fruits can either be eaten as a great snack, added to mixes such as pancake or muffin, or rehydrated and eaten soft and juicy.

Don't have a dehydrator yet? While you shop or build, you may want to consider dried fruit from the store. While quite expensive, it is readily available and there are good choices: apples, prunes, raisins, cranberries, strawberries, apricots, pineapple, and more. The down side is that most are heavily laden with sugar, but they are light and tasty.

Jello and instant pudding mixes are another long-term storage goody. Lasting indefinitely, they make a great snack, treat, or reward.

Dried beverages, whether they be coffee, tea, or powdered drink mixes, all store well, even in very long-term plans. It is best, as in all the other above items, to rotate your stock, because powdered drinks, especially, have a tendency to cake. Of course they are still usable, but it doesn't take much to use the old stuff as you go and replace it with new.

Nuts and sealed packages of sunflower seeds make another great storage snack. They will usually last several years, factory sealed or home canned. Otherwise, they will become as rancid as those opened holiday salted nuts. I can a variety of nuts at home, especially walnuts and pecans from friends' orchards.

Search stores and preparedness catalogs for other snacks that sound good to you. A person can always experiment (before spending money stocking up on an item) with just about any food.

MREs? For those of you who are uninitiated, MRE stands for "meals ready to eat," a meal in a pouch developed for the military, with no cooking necessary. Here I'll put myself on the firing line and say they just plain cost too much for this frugal person. They taste fairly good—about like a TV dinner—are reasonably nutritious, are certainly fast and easy to grab and run with, but they are expensive and heavy if you have to carry them.

However, their shelf life is quite good. It is claimed that they will store for five years, but I'd suspect quite a bit longer if kept away from heat. But, for the cost of an MRE to feed one person, I can fix a meal—a real meal where you get filled up—for four people, even in the boonies.

So there you have it—the truth according to Jackie on long-term food storage. Try it yourself and find out how creative your family can be. Mine certainly is.

Just remember these tips:

- 1. Keep food cool, dark, and dry.
- 2. Make sure the food is factory or home-sealed as well as it can be.
- 3. Rotate all storage food regularly, marking the date on which you entered each item into the pantry. Use the oldest first.
- 4. Don't be afraid to experiment.
- 5. Have fun.

After all, it's a real joy and very reassuring to know that your family can get by nearly any period of bad times, eating good, nutritious food that they enjoy. Δ

Chapter 5
Seven mistakes of food storage

By Vicki Tate

If you are going to store food, make sure that the food you store is adequate for the need you and your family anticipate. This may not be as easy to achieve as many people think, because the facts are that most people make serious errors when storing food—errors that will come back to haunt them when the food they've stored is the only thing that stands between them and their empty, dissatisfied, bellies.

There are seven common mistakes people make when storing food.

1. Variety

Most people don't have enough variety in their storage. 95% of the people I've worked with have only stored four basic items: wheat, milk, honey, and salt. Statistics show most of us won't survive on such a diet for several reasons. a) Many people are allergic to wheat and may not be aware of it until they are eating it meal after meal. b) Wheat is too harsh for young children. They can tolerate it in small amounts but not as their main staple. c) We get tired of eating the same foods over and over and many times prefer to not eat, then to sample that particular food again. This is called appetite fatigue. Young children and older people are particularly susceptible to it. Store less wheat than is generally suggested and put the difference into a variety of other grains, particularly ones your family likes to eat. Also store a variety of beans, as this will add color, texture, and flavor. Variety is the key to a successful storage program. It is essential that you store flavorings such as tomato, bouillon, cheese, and onion.

Also, include a good supply of the spices you like to cook with. These flavorings and spices allow you to do many creative things with your grains and beans. Without them you are severely limited. One of the best suggestions I can give you is buy a good food storage cookbook, go through it, and see what your family would really eat. Notice the ingredients as you do it. This will help you more than anything else to know what items to store.

2. Extended staples

Never put all your eggs in one basket. Store dehydrated and/or freeze dried foods as well as home canned and "store bought" canned goods. Make sure you add cooking oil, shortening, baking powder, soda, yeast, and powdered eggs. You can't cook even the most basic recipes without these items.

3. Vitamins

Vitamins are important, especially if you have children, since children do not store body reserves of nutrients as adults do. A good quality multi-vitamin and vitamin C are the most vital. Others might be added as your budget permits.

4. Quick and easy and "psychological foods"

Quick and easy foods help you through times when you are psychologically or physically unable to prepare your basic storage items. "No cook" foods such as freeze-dried are wonderful since they require little preparation, MREs (Meals Ready to Eat), such as many preparedness outlets carry, canned goods, etc. are also very good. "Psychological foods" are the goodies—Jello, pudding, candy, etc.—you should add to your storage. These may sound frivolous, but through the years I've talked with many people who have lived entirely on their storage for extended periods of time. Nearly all of them say these were the most helpful items in their storage to "normalize" their situations and make it more bearable. These are especially important if you have children.

5. Balance

Time and time again I've seen families buy all of their wheat, then buy all of another item and so on. Don't do that. It's important to keep well-balanced as you build your storage. Buy several items, rather than a large quantity of one item. If something happens and you have to live on your present storage, you'll fare much better having a one month supply of a variety of items than a year's supply of two or three items.

6. Containers

Always store your bulk foods in food storage containers. I have seen literally tons and tons of food thrown away because they were left in sacks, where they became highly susceptible to moisture, insects, and rodents. If you are using plastic buckets make sure they are lined with a food grade plastic liner available from companies that carry packaging supplies. Never use trash can liners as these are treated with pesticides. Don't stack them too high. In an earthquake they may topple, the lids pop open, or they may crack. A better container is the #10 tin can which most preparedness companies use when they package their foods.

7. Use your storage

In all the years I've worked with preparedness one of the biggest problems I've seen is people storing food and not knowing what to do with it. It's vital that you and your family become familiar with the things you are storing. You need to know how to prepare these foods. This is not something you want to have to learn under stress. Your family needs to be used to eating these foods. A stressful period is not a good time to totally change your diet. Get a good food storage cookbook and learn to use these foods! It's better to find out the mistakes you'll make now while there's still time to make corrections.

It's easy to take basic food storage and add the essentials that make it tasty, and it needs to be done. As I did the research for my cookbook, *Cooking with Home Storage*, I wanted to include recipes that gave help to families no matter what they had stored. As I put the material together it was fascinating to discover what the pioneers ate compared to the types of things we store. If you have stored only the basics, there's very little you can do with it. By adding even just a few things, it greatly increases your options, and the prospect of your family surviving on it. As I studied how the pioneers lived and ate, my whole feeling for food storage changed. I realized our storage is what most of the world has always lived on. If it's put together the right way we are returning to good basic food with a few goodies thrown in.

(Cooking with Home Storage is available from *Backwoods Home Magazine* for $16.95 plus $3 S&H. Send check or money order to *BHM*, P.O. Box 712, Gold Beach, OR 97444 or call 1-800-835-2418.) Δ

Chapter 6
Cooking from long-term food storage

By Jackie Clay

You should have at least a year's worth of food and essential supplies stored up in a large pantry. Unfortunately, actually eating from long-term food storage conjures up images of consuming endless tedious, tasteless meals of boiled rice and beans. You know—"survival" food.

But this is not how my household works. If I were to serve such meals, there would be total rebellion. After all, we've had at least two years' food stored for 20 years, and we eat daily from this food. We are not martyrs, and we do not eat tasteless food in order to be "healthy." Instead, we eat three meals a day from good tasting, comforting, healthy food, much of it home-raised, home-grown or harvested from the wild.

With a little practice and imagination, any family can quickly learn to produce meals, drawn from your food storage, that draw rave reviews from family members. And you do not need to spend hours in food preparation time either.

Buying for long-term food storage

A lot of problems arise when a family purchases foods they are not used to eating, and are not especially fond of in the first place. During hard times, or in an emergency, are definitely *not* times to begin eating such "survival" food. When you're stressed out, worried, and depressed, you need a lift, not countless meals of boring, tasteless food.

Take stock of the foods your family really likes. Then build your storage pantry on these preferences. Just about anything you regularly eat can be included. One notable exception is hamburgers and cheeseburgers. You can home-can hamburgers, by lightly frying tiny patties, then stacking them into wide mouth jars. I do a few, just for novelty, layered with finely chopped onion. When ready to use, refry them, adding cheese if desired. They're

Old-fashioned baked beans

2 cups dry navy beans
¼ cup ham or bacon flavored TVPs or canned ham
½ cup dehydrated chopped onions
8 Tbsp. molasses
4 Tbsp. honey
1 Tbsp. dry mustard
3 Tbsp. vinegar
½ cup tomato sauce
¼ cup catsup

Sort beans, soak overnight in water enough to cover. In the morning, drain beans, discarding water. Place beans in 6-quart or larger heavy pot with 12 cups water and simmer, covered for just long enough to get beans tender (older beans require longer cooking). Drain and discard water. In a 3-quart or larger casserole, mix beans with other ingredients and bake at 350 degrees for 1½ to 2 hours, adding water if necessary to keep beans from drying out. Serve hot with fresh whole wheat bread for a comforting, hearty meal.

good, but aren't the same as the regular hamburgers the family is used to. And there isn't an alternative to home canned hamburger available for purchase.

Have your family go through the supermarket with you and take notes as to what interests them, even if you have not cooked it before, or recently. You need variety in your long-term storage foods. Meals are like buildings. They need foundation blocks like potatoes, rice or beans, but also windows and doorways like fruits, meats, vegetables, and spices.

Choose the foods carefully, taking into consideration those you use daily or would use, if you had the time....and had no alternatives. Two frequently overlooked items are shortening/margarine and eggs. And make sure there are plenty of opportunities for "goodies," such as cookies, pies, puddings, Jello, etc.

Introducing the family to seldom-used staples

There's a definite place in a long-term storage pantry for such staples as dried beans and rice, but instead of buying all navy beans and white rice, consider that there are over a dozen varieties of beans and perhaps six varieties of rice available, each with its own unique taste, texture, appearance, and uses. In our pantry, I have many kinds of beans which we use: navy, red kidney, pinto, Anasazzi, black, and several varieties of Native American beans that we grow at home. You might include a couple of limas (which our family just does not like), cowpeas, Great Northerns, or others.

Then try a few really good recipes and gently ask all family members to try just one small serving with a meal they love.

Work your way into beans. Don't just cook up a huge pot and insist everyone eat them for a meal. You can "sneak" beans into meals by mashing cooked pintos in with taco filling, putting a few mashed beans in a layer of casserole, or into a hearty vegetable stew or soup.

Old-fashioned baked beans, complete with ham or bacon chunks, molasses and catsup will usually do the trick on a cold winter's day.

The same with rice. Few people like a big plate full of steamed white rice. "I don't like rice!" is the usual comment. My answer is get to know rice and all its uses.

Perhaps the easiest way to interest a family member in rice is to serve Oriental fried rice, complete with little chunks of chicken or pork. Spanish rice is another interesting way of introducing rice into a meal. The point is to introduce staples gently into meals, *before* called on to use them every day.

Oriental chicken fried rice

2 cups cooked white rice, cooled
½ cup finely diced onions (you may substitute rehydrated dry)
¼ cup rehydrated freeze dried or air dried green peas
1 cup rehydrated shredded carrots (or fresh)
¼ cup oil or shortening
1 Tbsp. peanut butter
½ cup chopped cooked chicken
2 Tbsp. soy sauce
2 eggs, equivalent in dehydrated egg powder
Spices to taste, including garlic, turmeric, hot pepper

In a large, heavy frying pan, heat oil. Add rice, onions, chicken, and carrots. Stir frequently with spatula until rice begins to lightly brown. Add peanut butter (no, it doesn't taste "weird"), soy sauce, peas, and spices. Continue stirring while flavors mix. As rice mixture appears to be done, quickly add beaten egg mixture and continue stirring with spatula until egg is cooked. Serve at once with soy sauce, sweet and sour sauce, or hot mustard sauce.

The use of ethnic meals is a great way to introduce a family to basics, as most less-wealthy nations have fabulous recipes, using cheaper staples as the base for meals. You can learn to do wonders with corn flour and corn meal, beans and chiles, rice, vegetables, sauces, and bits of meat, potatoes, and vegetables. I quickly learned that America is one of the few nations in the world where meat is used as the foundation of a meal. (We had adopted two older children from India and three from Korea, as well as sponsoring a family of nine from Vietnam.)

When cooking from a long-term storage pantry, learn to buy or home-can meat in smaller cans and jars. This allows one to use the meat as flavoring and texture to a meal, making a little go a long way, and not give the impression of "making do" or "surviving."

Homemade pizza

Crust:

2 cups flour
½ tsp. seasoning salt
¼ cup olive oil
2 tsp. dry yeast
1 cup (plus) warm water

Mix dry yeast and one cup warm water. While softening, mix other ingredients in medium bowl. Add softened yeast and enough warm water to make a soft, but not tacky, ball of dough. Work dough with hands, greased with a small bit of olive oil until elastic, then set aside in bowl, covered, for half an hour to rise. Oil baking pan with liberal olive oil, press out dough with hands. Prick dough with fork every few inches to avoid bubbles. Bake at 350 degrees until just barely done. It will not be browned but will lift easily from pan when picked up with a fork at corner.

Topping:

1 cup thick tomato sauce
1 tsp. brown sugar
1 Tbsp. rehydrated green peppers
1 Tbsp. dry onion
¼ cup chopped olives
¼ cup sausage TVPs
1 tsp. oregano
¼ tsp. garlic powder
½ tsp. basil
1 cup mozzarella or
½ cup dried grated parmesan cheese

Spread tomato sauce evenly on baked crust and sprinkle brown sugar, green peppers, olives, and spices on top. Top with cheese. Bake until cheese is barely golden brown and bubbly. This is yummy, and it is much in demand at our house. As a bonus, there are many variations including making a double batch of crust, using the second dough to be formed as bread sticks, which can be brushed with tomato sauce, herbs, and sprinkled with cheese and baked at the same time as the pizza. The bread sticks, dipped in a warm herbed tomato sauce, make a great addition to the steaming pizza.

A couple of examples that we regularly use are Oriental Chicken Fried Rice and Tamale Pie. I *never* have leftovers.

Such goodies as homemade pizza are always hits at home, especially in an emergency or hard-times situation. Key ingredients, such as pepperoni and mozzarella cheese, are in few long-term storage pantries. When one has their own dairy animal, the cheese is a snap, and dry-cured pepperoni lasts for months without refrigeration under cool, dry conditions. I also have canned chunks of pepperoni and am going to try mozzarella cheese too. But without these options, pizza is still a definite "go". Check out recipe above.

But we'll miss fried foods!

I think one of the things a family misses most, living totally from a long-term storage pantry, is simple fried foods. Now this can have an up side, as well as a down. The fewer the fried foods consumed, the better health we enjoy. But, honestly speaking, some fried foods do a lot to boost our morale during rough times.

Now, of course, if a family has their own garden, which a self-reliant family should have anyway, they will have abundant potatoes, fresh or in the cellar. If not, you can fry up a batch of canned potatoes, from time to time, or make potato patties out of leftover mashed potatoes (adding two beaten eggs to hold them together). There are also dehydrated and freeze-dried hashbrowns that are quite good.

Fried tuna patties

2 cans light tuna, drained
1 cup crushed dried bread crumbs
¼ cup dehydrated onion flakes
3 eggs, rehydrated equivalent
½ tsp. lemon pepper
flour to coat patties
oil to fry

Mix drained tuna, bread crumbs, egg, onion flakes, and lemon pepper. Divide into golf ball-sized portions, pat into patties, dip both sides in flour. Heat oil to medium heat and gently place patties into frying pan. After one side is done, turn and finish cooking. This is a quick and easy alternative to "fish sticks."

One satisfying, simple recipe we enjoy from our pantry is fried tuna patties. This provides meat, as well as satisfying an occasional craving for "fried food."

Roast beef hash is another "alternative" fried food. Unless fresh meat, either home grown or wild, is available, there will be little fried meat available.

One of the home storage pantry's best capabilities is providing quick, nutritious soups and stews at quick notice. These include those using pasta and noodles, as well as the more traditional. One of our favorites is homemade noodles, cooked with chicken broth. You can use store-bought noodles, but there is absolutely no comparison in taste or texture. And noodles are very easy to make, only taking a few minutes once you get the hang of it. Even "mistakes" are very edible.

Tamale pie

½ cup cornmeal

½ cup white flour or freshly ground whole wheat flour

¼ cup honey or white sugar

¼ cup shortening

1 egg (equivalent in dry egg powder, rehydrated)

1 cup rehydrated dry milk (+or -)

2 tsp. baking powder

1 tsp. salt

½ cup cooked hamburger (I use home canned) or beef TVP

2 cups tomato sauce

½ cup dry chopped onions

¼ cup dry chopped green peppers or chile peppers

2 tsp. mild chile powder

¼ cup dry sweet corn or ½ cup canned corn

Mix first eight ingredients well, making a medium batter (not runny or not stiff). Then in medium sized cast iron frying pan or 8 x 8 cake pan, mix the last six ingredients well, then top with cornmeal batter. Bake at 350 degrees until top turns golden brown. Serve hot with cold salsa.

Breads

Don't forget the staff of life. Breads provide an endless base to home meals, unlike their tasteless plastic wrapped cousins from the store. On one camping trip, we made an entire meal out of a crispy, fragrant loaf of French bread, without a dab of butter.

Breads can be made of varying flours for entirely different tastes, textures, and appearances. Aside from the "normal" white flour from the store, one can, and should, grind their own grains, producing a wide, wonderful, array of fresh flours. This produces a wholesome taste that most folks have never even dreamed of.

As all grains store much longer as whole grains, it is wise to stock up on these grains and grind the flours as you need them. Some suggestions are red and golden hard wheat for bread, soft wheat for pastries, flour corn for cornmeal, hominy corn for corn flour, buckwheat for pancake flour, rye for rye and pumpernickel bread, and rice for Asian cooking.

You will find that when you are cooking solely from the long-term storage pantry under times of duress, just the milling of the flour and baking of bread will bring peace and contentment to the whole family.

While "plain" bread will probably be most often used, stretch your creativity by expanding to more "exotic" breads such as pitas, tortillas, and sweet rolls, both for taste and variety. Most use about the same simple ingredients, and with just a little variation you can create a whole spectrum of tastes and possibilities.

One of our favorite breads is a versatile quick roll recipe. This recipe makes soft, tender dinner rolls, but also free-form breads, hamburger buns, sweet rolls, and coffee cake with little change.

Roast beef hash

1 pint (16 oz) canned roast beef (or wild meat)
½ cup rehydrated onion flakes
1 quart (32 oz) canned potatoes, drained well
oil to fry

Grind meat, potatoes and onion together with hand meat grinder. Heat oil to medium heat in large frying pan. Slide hash into pan, being careful not to spatter. Arrange the hash in a shallow layer, covering the bottom of the frying pan. Allow to cook, turning and stirring with spatula. Add seasonings, finish frying to preference, and serve. Popular condiments include salsa and catsup. Hash makes a satisfying one-dish meal. Leftovers are great for breakfast with scrambled eggs.

Tips for cooking from home storage

Cooking from a long-term storage pantry is easy, basic, and very fulfilling. But it is not something one learns to do overnight. It's sort of like gardening. The time to learn to garden is *not* when the trucks stop hauling food to supermarkets and there are acute food shortages. It takes time to get into the rhythm of gardening, learning what works, what does not, and how to do the most work with the least effort.

When a person plunges into cooking solely from home storage, they are quite often frustrated by "all the hard work," the poor results from their cooking, and the lack of enthusiasm from family members. Sort of like when a new young bride begins cooking for her husband for the first time. The results are often ho-hum.

Try taking one day a week, at first, to practice cooking out of your storage pantry. Ease into it with a few of these recipes; they are easy and basic. Then expand to others, found in some of the books listed below. Mennonite, Amish, Mormon, and Seventh Day Adventist cookbooks usually provide a good start for comforting, tasty meals with basics, as these religions stress commonsense preparedness and good family eating from basic, healthy ingredients.

It's well to mention at this time that it makes good sense to develop a small garden, at least, and learn to forage for wild foods (which are great tasting, by the way), while learning to cook meals from the food you have stored. Fresh foods are a very welcome change to dehydrated and canned, and they not only taste great, but provide extra nutrition, which could conceivably be lacking in a few long-term storage foods.

Likewise, if it is at all possible, develop your own source of fresh milk, eggs, and meat. Grandmas all over the world raised a small flock of chickens, even in town. It's funny that the U.S. is one of the few countries where this is not common today. Remember that variety is truly the spice of life; cultivate all the variety you can in your family's diet.

Halftime spoon rolls

Dissolve 2 tsp. dry yeast in ½ cup warm water, and set aside. Combine 1/3 cup shortening, ¼ cup of sugar or honey, 1 tsp. salt with ¾ cup hot milk (reconstituted dry or fresh). Cool to lukewarm by adding ½ cup cold water. Add 1 egg (or equivalent reconstituted dry) and softened yeast. Mix in 3½ cups sifted flour. Cover in same bowl, letting rise in a warm place till doubled. Stir dough with greased spoon. With an ice cream scoop (works easiest) dip sticky batter into greased muffin tins, filling half full. Let rise and bake at 375 degrees until golden brown. Remove from oven and brush margarine on top to soften nicely. I promise raves from this one. And you can modify it easily. With the addition of just a little more flour (about ½ cup), the dough will be firm enough to handle lightly, which makes forming hamburger buns on a greased cookie sheet easy, or forming into caramel rolls, cinnamon rolls, or coffee cake.

Basic whole wheat bread

Heat 4 cups milk (rehydrated dry or fresh). Soften 4 Tbsp. dry yeast in ¾ cup warm water. Add ¼ c. honey to warm milk, along with 1/3 cup oil or melted shortening, 2 tsp. salt, and 2 eggs.

Beat well, then as milk cools to luke warm, gently add yeast. Add about 14 c. fresh whole wheat flour, one cup at a time, mixing after each. When nearly stiff, mix with hands, incorporating just enough flour to make an elastic, workable ball of dough. Don't stop if it's sticky, and don't get it too stiff. Knead on a floured board for 10 minutes. Grease a large mixing bowl, place ball into bowl, and grease top. Pull ball out and put back in, greased top up. Cover with a warm, damp kitchen towel and let rise in warm place until about double. Knock down and let rise again. Divide into two or three loaves and place into greased bread pans. Preheat oven to 350 degrees. When loaves are nearly doubled in size, place in oven and bake for about 35 minutes until tops are golden brown. Grease tops with margarine to soften. Enjoy the best bread you've ever tasted.

Homemade noodles in chicken broth

1½ cup flour (either freshly ground whole wheat or white
¼ tsp. salt
2 eggs, reconstituted or fresh

Place flour in mound on board, making a nest or well in the center of the mound. Pour eggs into nest. Beat the eggs with a fork, gradually bringing the flour into the mix. Work the dough into a ball with your hands, picking up only as much flour as it takes to make a stiff, but workable ball. Knead the dough for about five minutes. It should not stick to the board. If it seems too moist, add a little more flour; if too dry, dampen your hands and knead longer.

Divide the ball into quarters. Cover three and reserve one to work with immediately. Lightly sprinkle board with flour and roll out dough, pulling it into a uniform thickness oval. Make it as thin as workable and let rest in a warm, dry place. Repeat with other three quarters.

When all dough is dry, but not stiff and brittle, roll like a jelly roll, cutting into desired thickness with a sharp knife. You can then either fluff out to separate and then carefully hang to dry or lay it flat to air dry for an hour.

Pour a quart of chicken broth (or use dry chicken granules to make a broth) into a large pot. Add diced, canned, or freeze dried chicken meat, if desired, as well as onion, carrots, and spices as wanted. Bring to a medium boil, then carefully add noodles, simmering just long enough to make them tender. The flour on the noodles provides natural thickening. You'll get raves for this simple, yet satisfying meal.

You'll find you get into the rhythm of this type of cooking easily. And you'll quickly develop time-saving ways of doing things. For instance, it didn't take me long to decide that if I ground a week's worth of grains I truly saved time, and cleaning of the grain mill, and I always had the grain I needed on hand conveniently.

I can honestly say that it takes only minutes longer per meal to prepare a great dining experience from my pantry than it does to rip and pry plastic wrap from something that appears to be food. And it makes the whole family feel great. We believe in *living*, not just surviving! Δ

Chapter 7
Storing water for an emergency

By Vicki Tate

Any of us who've thought much about emergency preparedness realize that one of the most critical items to store is water. Without sufficient water to see you through an emergency that lasts more than a few days, you and your family are at great risk. You simply can't live without water.

Fourteen gallons of water per person is the suggested amount to store for a two-week emergency situation. This amount is enough for subsistence purposes only: two quarts for drinking and two quarts for cleaning and bathing purposes a day. When you consider that a person normally uses in excess of 140 gallons of water per day for drinking, bathing, laundry, dishes, watering lawns, etc., this isn't a lot of water. If you have the room to store more you will want to do so.

The easiest way to store the bulk of your water is in 55-gallon polyethylene (plastic) water drums. These can be obtained from most food storage companies or from local container companies found in the yellow pages. It is important that you use only *food grade*, good quality containers. Many times you can get food grade containers from companies that distribute beverages or syrups. If you clean them well, they can provide a good container that costs considerably less.

One word of caution: often the taste or odor of the previous contents has leached into the plastic and over time may be reintroduced to your water. If you plan to use previously used containers, make sure that what it had in it before is something you wouldn't mind tasting or smelling in your water.

Most water containers come in 5-gallon, 15-gallon, or 55-gallon sizes. I always suggest that a family stores between two and six of these smaller containers, along with their 55-gallon drums. This is a prudent suggestion in situations where you might need to transport water in the normal course of events or in a situation where your normal water source might be disrupted, such as after an earthquake, hurricane, etc., and you might have to go to a secondary water source such as a water truck, stream, etc. to refill.

Water weighs approximately 8 pounds per gallon, so 55-gallon drums are much too heavy to handle (440 pounds), plus they are awkward. Smaller containers don't hold enough water and would require too many trips, especially if you have to go to a source on foot. Five to fifteen-gallon containers are more practical and can easily be put into a wheelbarrow or child's wagon and wheeled to and from an area.

Two-liter pop bottles make a good container for additional water storage and cost nothing if you save them and fill them with water as you empty them. To economize many people are tempted to use empty milk jugs, but don't plan to store water in these for more than three to four months. They are biodegradable and will break down within six months. Not only may you lose your water, but if they are stored near food or other items they may damage them. Heavy containers should always be stored close to ground level and secured to prevent breakage or possible injury in the event of earthquake, etc. Be sure to store your water away from any harmful chemicals or objectionable-smelling products.

Culinary water (tap water) is what is usually stored for long term storage. If you have a clean, opaque container where the light cannot get through and your water is bacteria-free when you store it you probably don't need to treat it further. Under these conditions the water actually gets more pure as it is stored. However, for most of us there is no guarantee that our culinary water is bacteria-free, so most of us prefer to treat our water in some way as a precaution as we store it. Several methods have traditionally been used to purify water for long term water storage.

> 2% Tincture of iodine—To use this add 12 drops per gallon of water. Note: pregnant or nursing women or people with thyroid problems should not drink water with iodine.
>
> Chlorine bleach—Household bleach can also be used. This should contain a 5.25% solution of sodium hypochlorite without soap additives or phosphates. Use 1/8 teaspoon (about 5-8 drops) per gallon of water.

Most of us have used one of these methods to treat our water over the years. Both are inexpensive and are effective methods of killing bacteria. I have always preferred the iodine method myself. The one drawback, however, is that both may have negative health effects if used for long periods of time.

I was introduced to a product a couple of years ago that I now prefer to use instead because it is an excellent water purifier, but it also has many medicinal properties. It is a stabilized oxygen called Ion that is effective in killing all harmful bacteria without any of the harmful health effects associated with chlorine or iodine. For long term storage add 20 drops of Ion per gallon of water. One bottle will purify two 55-gallon drums. It is also excellent for your emergency packs (72-hour kits). It is small and light weight (2.33 oz.) but extremely effective.

Studies show that if water is bacteria-free and is stored in clean containers it will stay safe for several years. It is a good idea, however, to periodically check your water for purity and taste. And every few years it's a good idea to change it. One of the things that affects the taste of water is it "going flat." This occurs because of the oxidation that takes place as it sits. You can improve the taste by pouring the water back and forth between containers to aerate it or by beating it with a hand egg beater. You also may want to store some flavorings such as fruit drink powders, kool-aid, etc. to add to your water if you find the taste objectionable. One of the other benefits of Ion is the oxygen remains suspended, maintaining its good taste for much longer periods of time.

Remember also that you have several sources of water already in your home that can be tapped in an emergency, such as your hot water heater, toilet tanks (don't use water from a tank that contains colored disinfectant, as it is poisonous), water pipes, ice in the freezer, etc.

Water is relatively inexpensive to store and certainly not difficult to do, but certainly the time to store it is now. We take water for granted when things are normal, but in an emergency it becomes absolutely critical. This is an item you can't afford to overlook in your preparedness preparations. Δ

Chapter 8
What if the electricity goes off?

By Michael Hackleman

The name changes, the date shifts, but it's the same question: Are you ready? Earthquake, fire, flood, plague, meteor strike, nuclear attack, hurricane, tornado, and terrorist attack—all strike in the moment. Economic collapse, crop failure, famine, and nuclear winter are forces of siege that could last months or years.

From a distance, the first evidence may be a blackout or a news report. The area affected by the disaster will dictate the probability, frequency, and durations of blackouts. If the scope of the disaster is large, other services—water, natural gas, gasoline, fuels, food, and goods—will fail. Following that will be the loss of phones, police, fire, rescue, utility, Red Cross, and government services.

It is said that crisis has two components: danger and opportunity. There is danger in a crisis—catastrophe, collapse, and chaos. There is opportunity in crisis—restoration, renewal, and revival. Preparedness doesn't mean you'll survive, but it won't contribute to your demise.

There are issues that are specific to living in a city or in the country, so I will treat each as distinct scenarios. As you discover issues that approximate ones you may experience, you will likely be drawn to research these topics in more detail.

A blackout is a likely scenario in either a short-term or long-term crisis, so that is a good place to begin.

Blackout ready

The pressing question when a blackout occurs is: When will it end? Virtually anyone can put up with a few hours of interruption. Just break out the candles, don warmer clothes, and read a book or enjoy the company of a friend. The average blackout is a pop quiz. "Are you ready?" it asks. When the blackout continues, with no end in sight, the need for light, heat, water, and food grows.

There are four critical loads in a home affected by a blackout: lighting, heating, refrigeration, and the water system. More specifically:

• **Lighting**. Lighting is essential for overall safety, particularly at night. Fortunately, it need not be electric. Candles, flashlights, and kerosene lanterns are traditional lighting sources for blackouts. Preparation for a blackout requires stockpiling matches, candles, batteries, or fuel for lanterns. Don't forget to put this stuff where you can find it in the dark!

• **Heating**. Central air heating systems, even if they use natural gas or propane, depend on electricity for the blower that will circulate the heated air. During a blackout, this system will not work. Areas with temperate climates allow most users to compensate with warmer clothing and the use of small propane or kerosene heaters. Wood stoves are also a popular alternative to central heating systems.

• **Refrigeration**. A refrigerator will keep things cool for a long time after power is interrupted. From the beginning, minimize the frequency and duration of opening its door to preserve its cool! As the blackout continues, consume the more perishable items first. Even a small stockpile of canned or freeze-dried foods will prove helpful during a blackout. Unless you've arranged for a way to heat and cook food, ensure that your supply is edible "as is," or with simple re-hydrating with water.

• **Water system**. Most community water systems are designed to work for some time following a blackout, powered with huge standby generators. Private water systems built around streams, springs, and wells that use electric pumps will quit working as soon as the electricity goes off. The pressure tank will still deliver some water, so immediately fill handy containers (bottles, buckets, bowls, bathtub, etc.) before this supply is depleted. The standard household water heater is another source of 30-50 gallons of water. How will you handle toilet, shower, and sink during a blackout? Some forethought and planning will help with these processes during an extended blackout.

Other sources of electricity

Utility electricity available at the wall socket in a home or business is rated 120 Volts and 60-cycle AC. There are two ways to supply this same specialized power in a blackout: a standby generator and a battery-powered inverter.

The standby generator:

Where the interruption of utility power even for a few hours is critical—i.e., emergency equipment and services in businesses and hospitals—a standby generator may be used to supply power. A standby generator is an engine combined with a generator. This unit may be started manually or automatically and requires only fuel (gasoline, diesel, or propane) to operate until grid power is restored.

Homes may also use a standby generator to supply electricity during a blackout. A common arrangement is to start the backup generator from a remote control panel in the house. Some or all of the household circuitry is transferred from the utility line to the standby generator. This process is reversed when utility service is restored.

A standby generator for homes, businesses, or hospitals is usually rated to handle only some of the existing loads. A generator large enough to handle *all* of the loads is big and expensive to buy, maintain, and operate. A detailed analysis of existing loads should precede the installation of any standby generator. Make a load list. This is a good place to rate loads as essential or nonessential. Later, this helps identify circuits that will be left ON or shut OFF during generator operation.

In theory, the standby generator seems like the best way to handle blackouts. However, there are five reasons why it is less than an ideal solution: expense, fuel supply, peripherals, efficiency, and sound.

• It is a fairly expensive system for only occasional use. For a big chunk of time, the generator is not doing anything for you at all. Standby generators designed for long life and minimal noise are more expensive than ones operating at higher rpm (3600 rpm).

• Requires fuel to run. Either you must install a large fuel tank nearby or you'll be transporting fuel cans to and from town to feed a rather thirsty beast. Weather severe enough to require generator operation is rarely the best time to travel to refill empty gas cans.

• Needs peripheral hardware to work. Remote startup. Transfer switch. Monitoring gauges. Fuel supply. A firesafe, weatherproof installation (shed?). A battery that is ready to start the generator. Add these costs to that of the generator itself.

• It is needed for even small loads. A generator powering a few loads has a much lighter load, but gobbles (inefficiently) fuel as though it's doing more work than it is. Either way, it experiences wear.

• It is noisy. This is a security issue. A standby generator lets everyone in the area know where you are. At the same time, proximity to the generator impairs one's own hearing. Bad combination.

Despite these limitations, standby generators have their place. There is 100 times more available energy stored in a pound of gasoline than a pound of battery. In the short term and for big loads, the generator gives the biggest bang for the buck. The questions are: how big is your need and what's the duration of the blackout?

The battery-powered inverter:

Another way to make electricity like that supplied by a utility is through an inverter. An inverter is an electronic device that converts DC electricity into AC electricity. (DC is direct current. AC is alternating current.) The result is identical to the stuff from the utilities, even cleaner.

One source of DC electricity is a battery. Thus, an inverter can transform the DC electricity from a battery into 120V, 60-cycle AC power. A battery is not truly a source of electricity. Rather, it is a means of storing the energy (in a chemical form) of the DC electricity supplied to it. The best sources of DC electricity are PV (Photo-Voltaic, or solar) modules, wind-electric machines, and small hydro-electric systems. More on this later.

A battery charger plugged into the utility line will also supply DC electricity to a battery. This is a popular idea. The batteries are charged and maintained at full readiness, and ready to substitute their energy for that of the utility for as long as they're able. The bigger the battery (bank of batteries), the longer the system can bridge the blackout.

These systems are common. Have you ever wondered why your phone works when a blackout occurs? Phones run on electricity, too. The phone company has a "standby" or backup system which switches ON automatically when utility power is interrupted. This is called an uninterruptible power system, or UPS. At the heart of the UPS system is a bank of batteries that are much like the battery in your car, except bigger and heavier. Those batteries store enough energy to run an entire complex of telephone-related equipment for many hours during a blackout. When a blackout lasts longer than that, an engine-driven generator (fueled by gasoline, diesel, or propane) is started up to handle the entire load and recharge the batteries.

Until a few years ago, a small UPS system was the primary way to avoid the loss of power to a computer during a blackout. A critical period is the time it takes to switch between utility and battery power. To avoid any glitch, early UPS systems would run the computer's inverter from the batteries full time, while utility electricity only maintained the battery pack's charge. Better electronics have improved the purity and speed of the transition time. Today, many computers are unaffected by the transition as newer "line-tie" inverters switch from utility to battery, or back, in milliseconds.

There are many applications where a split-second transition between utility and battery power is not an issue. Or where more modest loads are dictated. Here, a simple and less expensive system—a small inverter, battery, and transfer switch—works well. The system I installed when I lived in the city was sized to power a furnace blower (and its controls), refrigerator, stereo, and four lights during an emergency. The system was installed near the main distribution box (where the fuses or breakers are located). It involved moving the wires and breakers (to which the wires are connected) into a service box I added. With a transfer switch added between the two boxes, I could shift these three circuits between utility power or the inverter's output. This is basic electrical wiring, easy for a DIY (Do-It-Yourself) homeowner or a local electrician.

How did it work? I'd give the blackout 10 minutes before I looked for the load list. It's a map that lets me move about the house, shutting off unneeded loads on the few circuits that will be switched to the inverter. (The energy stored in the battery is the lifeblood of the backup system. Don't let it bleed away needlessly!) Next, I'd shut off the main utility switch, flip the transfer switch to inverter, and turn the inverter ON.

A home UPS system

The simplest backup system is composed of two major components—a battery and an inverter. Batteries and inverters come in all shapes and sizes. Select carefully and they will serve you well.

The important characteristics of a battery are voltage (V), capacity (Ah), and design cycle.

1. Voltage. Common battery voltages are 6-Volt and 12-volt (hereafter, 6V and 12V). Large battery banks will employ individual 2V cells with massive capacities.

2. Battery capacity. Battery capacity is rated in Ah (Amp-hours) or cold-cranking amps (useless for our purposes). The Ah rating is helpful in describing the amount of energy the battery can hold.

3. Design cycle. Batteries are of two types: SLI and deep cycle. (A cycle is a discharge and full recharge.) A SLI (starting-lighting-ignition) battery is used in a car to start the engine—a fairly shallow cycle—and is immediately recharged by the engine's generator. Deep cycle batteries are used in applications where the battery's energy may be nearly depleted—a deep cycle—in use. This process will damage a SLI battery internally and eventually result in its failure.

12V vs 6V batteries

The smallest deep-cycle battery you might use for an inverter is rated 12V and 110Ah. These are used in boats, trolling motors, and RVs. At 50-70 pounds, this battery is about as much as a healthy person can carry and maneuver in a confined space. To increase the capacity of this system, you "parallel" a second battery with the first (Fig. 1). To parallel a battery (same voltage only, please!), make connections positive-to-positive and negative-to-negative for both batteries and load. The voltage will stay the same; in this

Figure 1:

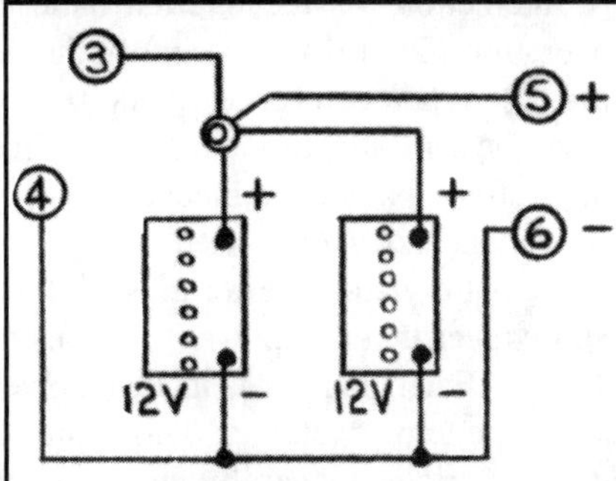

Batteries may be wired in parallel to increase capacity.

case, 12V. At any time later, you can increase the system's capacity (the rate or duration of power delivery) by adding batteries of the same voltage, even if they have different capacities.

A better building block in a battery bank for inverter operation is the 6V, 220Ah battery used in golf cars. It has half the voltage, yet twice the Ah of a 12V battery of the same size and weight. So, they have the same "energy density."

To supply the 12V electricity our inverter needs, two 6V batteries are connected in series (Fig. 2) like dry cells in a flashlight, with ONE wire connecting the positive of one battery to the negative of the other. (A novice may try to connect the other two posts together, which results in a very hazardous short-circuit.) The result of the series wiring is a new, bigger battery of 12V with the remaining posts, positive and negative, connected to the system in the same way as would be any 12V battery.

Theoretically, pound for pound, two 12V batteries in parallel will equal the capacity of two 6V batteries in series. In reality, a 6V battery is tougher—thicker plates, fewer cells to water, and greater tolerance to deep cycling and cold weather—than a 12V battery, resulting in a longer service life for almost any application.

Figure 2:

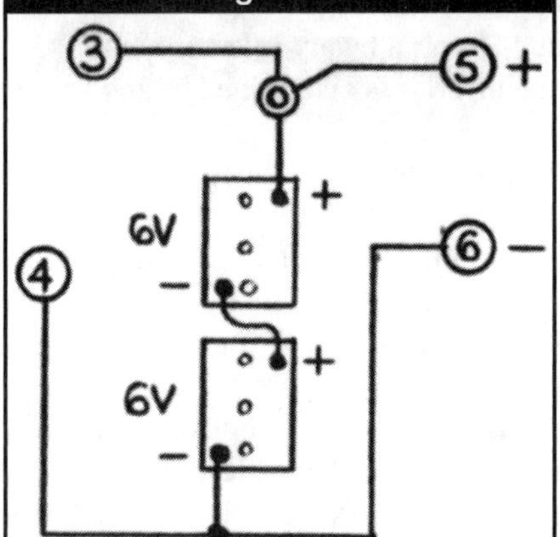

Batteries are wired in series to increase voltage.

Expect to pay $70-85 for a 6V, deep-cycle golf car battery (or equivalent). You'll need them in pairs for inverter operation at 12V.

Inverter features and ratings

Today's inverters serve two critical functions. First and primary, they convert the battery's output (low-voltage DC) to a form that your household can use (120 or 240 volts

AC, 60 cycles). Second—and most desirable for standby generator or utility interaction—is the internal battery charger option. A battery charger's operation (DC-to-AC) is simply the reverse of an inverter's operation (AC-to-DC). When combined in one box, the inverter and battery charger share (use) the same electronic hardware. In this way, utility electricity stores itself in a battery which, in a blackout, will release the energy, powering an inverter to make 120V, 60-cycle AC.

The battery and inverter must be "matched" to each other and to the loads you expect them to power. Appliances, lights, and tools are referred to as "loads." Each "load" has its own power (consumption) rating. You may have heard the term "wattage." This is an expression of the RATE at which a load uses electricity. Generally, lights and radios are small loads while refrigerators, motors, and toasters are big loads. The effect of loads is accumulative. That is, if you operate more than one load at one time, the total load is the addition of all those wattages. The power consumed by even one small light all night might be greater than that of a toaster operating for a few minutes.

In an emergency, you must reduce the loads the battery/inverter unit will power. The faster you use the energy stored in the batteries, the sooner you'll have a "second" blackout! Make sense? In a blackout, you become the power company, responsible for rationing both the rate and quantity of expected household needs for a specific time period.

Inverters have voltage and wattage ratings.

1. Voltage. The voltage ratings are divided into input and output. The input voltage is the DC voltage of the battery bank. Inverters exist to handle DC voltages of 12V, 24V, 32V, 48V, or 120V.

The output voltage of the inverter is the 60-cycle AC voltage. It may be 120V (commercial) or 220V (industrial), or both.

2. Wattage. Wattage ratings of inverters range from 50-4,000 Watts (4kW) or larger. What wattage works for you? Here's a handy rule-of-thumb.

a. The *minimum* wattage rating of the inverter is determined by the largest single load you expect it to power.

b. The *maximum* wattage rating of the inverter is the largest combination of loads you want it to power simultaneously.

For example, if you had loads of 50 watts, 120 watts, 220 watts, 1200 watts, and 1400 watts, the inverter rating could be as low as 1400 watts (for the biggest single load) or as high as 2940 watts (for *all* of these loads.)

High-power inverters are expensive and require more battery capacity. Smart owners balance this situation by avoiding simultaneous use of heavy loads. In this example, then, selecting a 2000-watt inverter would handle everything else if the operator avoids using the two biggest loads simultaneously.

The price tag of a small UPS system is well within the reach of many homeowners. Inverters average a dollar a watt and batteries (lead-acid type) about a dollar a pound. A battery/inverter system is virtually maintenance-free and tucks away on a shelf in the garage or carport, ready to work when the blackout comes. Fortunately, your investment in this system has a second success. It is the core of a system that enables you, when you're ready and able, to tap the renewable energy sources—solar, wind, and hydro—all around you.

A no-inverter DC system

Utility power, in the form of 120VAC, 60Hz, is very specialized power. In a blackout, you may have less need for it than you might think. It is well known that a car or truck is useful in emergencies for the radio, light, heat, and shelter it offers. Without the engine running, there is enough capacity in a vehicle's 12V battery to power lights, radio, and the horn for some time. Periodic engine startup adds heat to the equation and recharges the battery, too!

Similarly, a stand-alone 12V battery pack located in the garage or home may be kept on charge (with a battery charger) until utility power fails and its stored energy is needed "as is," at 12V. This does not mean that you can power the same 120V loads as an inverter will. The RV (recreational vehicle), automotive, and marine markets offer almost any type of appliance, motor, tool, pump, and light that will work directly from 12 volts DC. For example, several high-efficiency 12V fluorescent lights will provide 20-40 hours of welcome light from one automotive-size battery. I can think of nothing more reassuring in the darkness, particularly when a storm is raging, than the steady glow of a lamp.

How do you wire up a 12V system to be blackout-ready? For occasional use, clamp-type lamps and several lengths of extension cords may be connected together to distribute light through a dwelling. This assembly can be coiled up and put away until a blackout occurs. A more permanent solution is to dedicate an electric circuit to 12V use. Existing household circuitry rarely adapts easily to a dedicated usage (unless one is still building one's home). Here, a well-planned layout and one standard roll of Romex wire will add a 12V circuit to any home, shop, or building for lights and a radio.

Living beyond the grid

Most RE (renewable energy) systems are based around 6V and 12V storage batteries. The simplest RE systems use a solar module, one or two batteries, a few 12V lights, and a 12V radio. Except for the PV module, this is identical to the system (described above) to supply power during a blackout. Becoming blackout-ready, then, is a step in the direction of becoming energy-independent.

RE systems are generally located "beyond the grid." The cost of bringing in utility service even a mile is often more expensive than investing in a system that is utility-free. RE technology has focused on being modular. This makes it simple to add more capacity, and to move and re-install the system.

There are energy sources other than PV modules worthy of your attention: wind and water. Wind-electric machines and small hydro-electric turbines are also viable energy producers. A multi-source system is smart for three reasons:

1. Seasonally, wind and water sources of energy are complementary with solar-generated power.

2. Solar, wind, and water system hardware is designed to supply low-voltage DC, particularly 12V and 24V.

3. A system designed around one source readily accommodates additional sources. The systems are more similar than different. Therefore, the battery bank, distribution and fusing panels, and monitoring equipment are virtually the same and are shared by the different sources.

Putting together a backup or RE system is also a good way to learn the basics of electricity itself (i.e., volts, amps, watts, and amp-hours). I believe this is essential if one is going to rely on electricity for anything. With this knowledge comes an appreciation for how energy moves and changes, and how it can be harnessed to fill your needs.

Beyond a blackout

Preparing for something worse or longer than a normal blackout is a frightening prospect. I avoid being overwhelmed by the sheer immensity of the topic by dividing the issues into two phases: basics and preparation.

Basics:

Basics represent the checklist of life. What does a human being need to survive, short and long term? Air, water, shelter, and food.

Air: Few think much about breathing until they can't. Remedies that take longer than three minutes are of little value. Shelters must remain tied to the atmosphere directly. If there are airborne pollutants (smoke, ash, etc.), filters will be needed to breathe without risk of injury.

Water: Humans can live only three days without water. See that you store some or have access to it. Water is easily contaminated. Figure out a way to purify it. Drink and cook with pure water or risk illness.

Shelter: Human beings are amazingly manipulative of their environment, yet remain vulnerable to it in crisis. Shelter holds back the extremes of heat and cold, offers dryness, and feels safer for sleeping.

Energy: While we manipulate energy in our home, workplace, and car on many levels every day, it is all artificially generated. When that source is lost, the first job is to conserve it, in whatever form it is available. With any prolonged interruption of transportation or utility services in crisis, stockpiles of fuels like gasoline, diesel, and kerosene will be depleted or prohibitively expensive.

Food: Humans can live about a week without food, less in cold weather and limited water. In a mild emergency, stockpiling food, even enough for 5-7 days, saves having to forage, hunt, buy, barter, or trade for it. Or worse. Hunger strips away the resolve of people unaccustomed to its grip. Foodstuffs in most cities would disappear in a few days during a real crisis.

Preparation:

As one becomes more self-reliant, there is less dependence on (or need to buy) water, electricity, food, and fuels. Transportation needs also decline, allowing you more time to live and work at home. Coincidentally, this process prepares oneself for short and long-term disasters.

Figure 3:

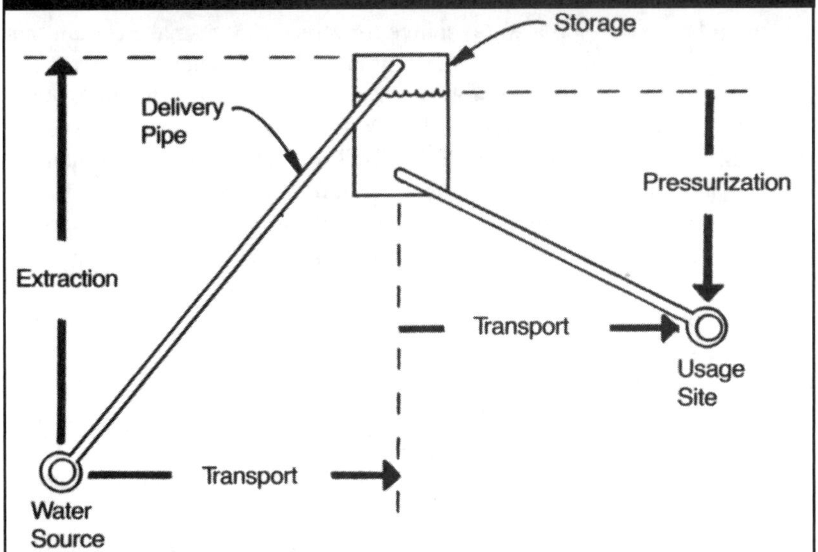

Water systems involve the processes of extraction, transport, storage, and pressurization.

Figure 4:

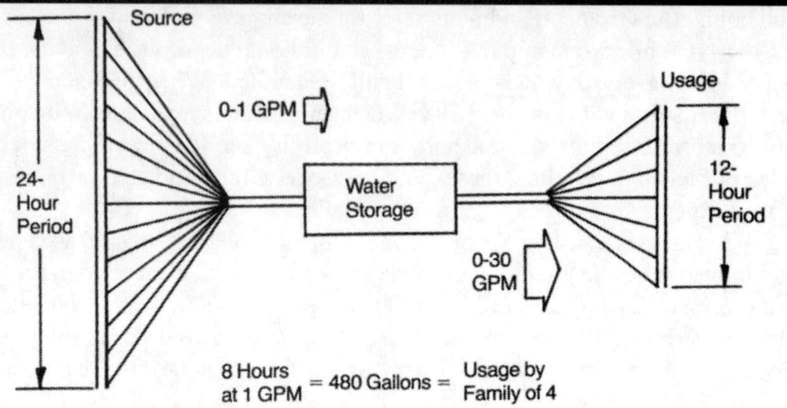

Low-yield sources can meet daily household water needs.

Here are some additional thoughts on preparations for water, food, energy, heat, lighting, motors, electronics, communication, and transportation.

Water: In a crisis, life is water. If your shelter—home, building, garage, cabin, RV, camper, tent, tipi, tarp, or cave—is connected to the town or city supply, your backup plan is to fill everything you can as soon as you can. If you can't develop your own source, prepare some way to store water in 5-55 gallon plastic containers, or plastic, wood, or steel tanks.

If you plan to develop a water source, or already have, make certain that your system is not completely dependent on utility or generator power. The process of water usage can be broken down into four areas: extraction, transport, storage, and pressurization (Fig. 3). Treat them as separate issues to maximize the versatility of the system. A low-yield water source quickly accumulates enough water to handle a standard household (Fig. 4).

There are low-power, low-voltage, and energy-efficient alternatives to the standard submersible pump. These can be piggybacked onto existing systems or work alone. PV modules powering a 12V or 24V pump (no battery) have seriously challenged wind-powered pumps in unattended operation, like livestock watering, in the past decade. Most renewable energy systems use something similar.

Water you waste also wastes the energy invested to get the water to you. A more active conservation method makes multiple use of the water. A "gray-water" system often doubles the usefulness of the water supply (Fig 5). Cooking, drinking and rinsing are the purest uses. Garden, clothes washing, and toilet are secondary uses. A plan and a bit of plumbing will help with this. There are several books on gray-water systems.

Figure 5:

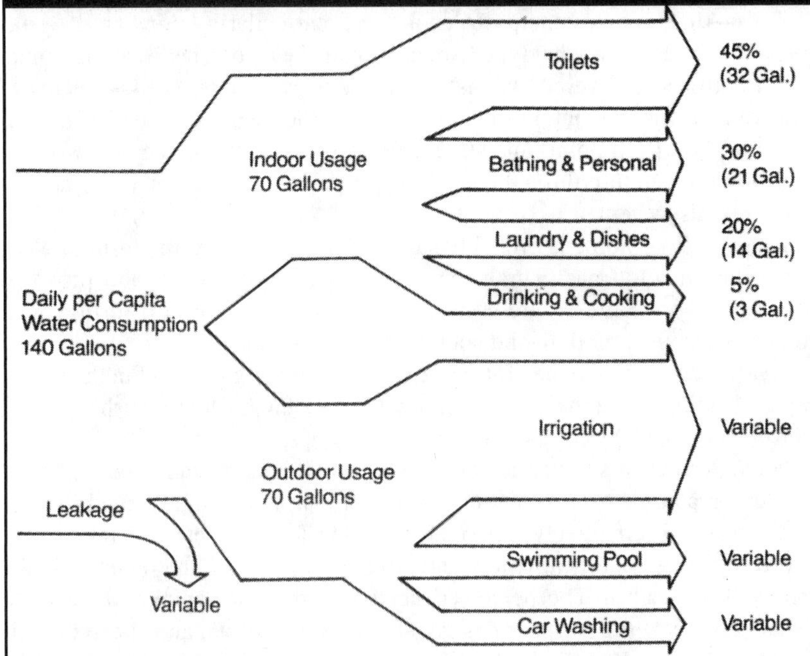

Understanding how water is used helps conservation and gray water system design.

Look at rainfall collection, cisterns, and pools as additional sources and storage methods.

Food: Food is one of the first concerns anyone will have in a crisis. Food issues revolve around supply, preservation, and cooking.

1. Supply is what you start with, if you don't grow your own. A stockpile, however small, is a good idea. Trading work or goods with people who farm and garden also works.

2. Preservation recognizes that food must be preserved against spoilage and infestation. Standard refrigerators and freezers work when there is abundant electricity. In an RE system, they hog energy. A high-efficiency, low-voltage refrigerator is expensive, yet rugged. More importantly, it frees up an appreciable chunk of energy that would be otherwise generated, stored, and inverted—only to be wasted. There are alternatives to refrigerators—canning and dehydrating, selective harvesting, and earth storage (i.e., a root cellar). Several good designs of solar dehydrators exist.

Using one or more of these techniques further reduces the load on, or the need for, a refrigerator.

3. Cooking. It takes energy to cook food, particularly grains and vegetables. How much? Of what type? Solar cookers are a good bet if you're home. A 24-hour solar-powered oven is possible. A parabolic tray of less than 100 square feet can heat natural oils in excess of 350 degrees F. (100 degrees F short of their flash point) and store a sufficient quantity to keep the oven of uniform temperature throughout a 24-hour period. Use gas or wood heat to back up this system.

Energy: Your home is probably supplied with energy in the form of electricity and natural gas. Rural homes may use wood energy and propane. These energy "sources" are converted into only a few useful forms: heat, light, mechanical motion, and sound (stereo and radio).

Heat: Heat is a cherished form of energy and the biggest load in the home. Space heating. Water heating. Cooking. Dishwashing. Clotheswashing. Both refrigerators and air conditioners are heat pumps.

Good designs of solar collectors exist to handle these heating tasks. While designing a home to use solar energy is optimal, many homes can be retrofitted to use it effectively. Thermal mass—water, concrete and rock—will store solar energy for nighttime and storms. Save wood and other fuels for really bad weather. The perceived need for air conditioning and massive heaters is a coverup for poor design, sloppy construction, and cheap materials. Good insulation is a must—floor, walls, and ceiling—to avoid heat loss in winter and heat gain in summer.

A good understanding of how heat moves (radiated, conducted, and convected) and what happens to radiated heat (transmitted, absorbed, and reflected) helps collect, contain, store, use, and release it.

Lighting: Lighting is essential for moving about at night, or in dark places. Still, night is for sleep, even in emergencies. Rest is important in survival. And sleeping saves light!

Incandescents, fluorescents, LEDs, and oil lamps all have value in lighting.

1. Incandescents, like standard household 120V bulbs and spotlights, gobble energy. Reserve their use to short durations. 12V automotive (turn signal type incandescent) bulbs are inexpensive, work directly on 12VDC, and are low-wattage. Still, use them sparingly.

2. Fluorescents, particularly those that are high-frequency (20KHz or above) are efficient and long lived.

3. LEDs are light-emitting diodes that operate at extremely low power. LEDs may be grouped together to increase voltage and light intensity. They're expensive but have a service life hundreds of times longer than incandescents.

4. Oil lamps will burn natural oils that may be pressed from many types of plants.

Motors: Motors convert electricity into mechanical motion. Motors power appliances in the home and tools in the shop. Pumps, fans, hair dryers, coffee grinders, juicers, turntables, tape decks, CD players, vacuum cleaners, computers, answering machines, and electric can openers use AC or DC motors.

High-wattage motors are difficult to power with low-voltage DC directly. Use an inverter or generator, as needed. Low-voltage DC motors may be substituted for AC ones under 2 HP. Or seek their 12V DC counterparts. Of course, manual tools don't need electricity to work.

Electronics: Electronic devices may be divided into two categories: high voltage and low voltage. The bigger and heavier the electronics, the more likely the need for 120V, 60-cycle AC. This includes the family stereo system, computers and peripherals, printers, television, and microwave ovens. Inverters and generators will be needed to power these units.

Light-duty electronics work around low-voltage DC, often below 12V. This includes remote phones, answering machines, portable radios, calculators, and portable CD and tape players. Look for a black module that plugs into the wall receptacle. The other end plugs into a DC input jack. DC input jacks may also be found on battery-powered units.

With a suitable DC-DC converter (or a dropping resistor), these electronic gizmos can be directly powered from a 12V car battery. (With a small modification, the jack can be re-wired to also recharge NiCads while they're in the radio.) Note the voltage printed near the jack to find the unit's voltage. Or count the number of cells (batteries) the unit contains and multiply by 1.5V to calculate the voltage. Or read the rating on the black module that plugs into the wall. This will help select the dropping resistor or converter setting.

Most electronic devices are polarity sensitive. By law, manufacturers are required to show the polarity of DC inputs, usually with a symbol. Wire the jacks and plugs accordingly.

Small 12V B&W TV sets may also prove handy, providing local coverage of a crisis. (Sorry, 12V color TVs gobble energy. Avoid using them.) These and other 12V devices often use a cigarette lighter plug (like the one that plugs into the car dash). If your vehicle doesn't have one, buy a lighter receptacle from an RV or renewables dealer. It can be clamped to the car battery posts or hardwired into the vehicle.

Communication: Details of what is happening beyond your own influence during a crisis is useful and, perhaps, crucial. In a blackout, the AM-FM radio in a car or truck may be the only communication at your disposal. At low volume, a radio will work for many days on just the car battery. You may need to position the vehicle (and antenna) away from buildings to get good reception. The news may not be reassuring if you're expecting help, but it will help you make better guesses or decisions about what you can and can't do.

Battery-powered, multi-band radios or boomboxes that use dry cells are equally good. With rechargeable cells (i.e., with NiCads) installed, there is no end to their useful service life. The cells can be recharged from renewable energy sources or even the 12V battery in a car. Note the actual voltage, use a converter or dropping resistor, and observe polarity. At low volume (or with earphones), these radios use only a tiny amount of energy in operation.

Transceivers, ham radio sets, walkie talkies, and CB (Citizen Band) radios are all useful, particularly for communities. Understand the power requirements to ensure that you can meet them. As well, recognize that sophisticated radio gear doesn't mean more effective communication. The semiconductor junctions in transistors and chips are extremely vulnerable to EMP (electromagnetic pulses) generated at high altitudes by both nuclear weapons and meteor strikes. The more complex something is, the more there is that can go wrong with it.

Transportation: Transportation may be adversely affected by crisis. Roads blocked with debris or other vehicles, bridges out, power lines down—these are common themes in a disaster. Owning a 4WD vehicle helps but it will need fuel, oil, tires, and parts to operate.

Vehicles converted to electric propulsion have an advantage over gas engines. There are only a few sources for gasoline. An electric vehicle (EV) is "fueled" by electricity from utility power, a standby generator, and renewable energy systems (solar, wind, or hydro). An EV has an additional advantage over vehicles with engines: it is silent in operation.

It may be easier to get around with motorcycles (noisy unless electric) and bicycles (mountain-type). Closer to home, carts, wagons, wheelbarrows, and garden carts will help with everyday work or emergencies. Again, with self-reliance, there is simply less need for transportation. Δ

Chapter 9
Medical kits for emergencies

By Jackie Clay

There may be a time, as close as tomorrow, when your loved ones need medication or medical treatment and there is no drug store open or doctor available. This may be as simple a situation as a head cold coming on during a weekend night, or more drastic, such as nothing available after a civil disturbance, terrorist attack, or natural disaster.

Family medical kits

Here at home, we've always had a medical kit. Several, in fact. One is quite large, made up of a poly box, originally designed as a field box for trap and skeet shooters. This "drug store on wheels" is a well-packed medical utility box that will handle nearly everything from a cold to severe lacerations. This one we carry when traveling in remote locations.

But while it is loaded with most medical needs, far surpassing a first-aid "kit," it is heavy, weighing over 30 pounds, and it is not something we carry for short trips.

An intermediate kit is lighter and fits into a flat, moderate-sized fishing tackle box. While this does not contain such a wide variety of medical supplies and medications, it is a very well thought out first-aid—and then some—medical kit. This is light enough to pack in our canoe (if we don't foresee many lengthy portages, when every ounce counts), with camp supplies on a horse packing trip, or small enough to take up little room in our truck.

Besides this kit, we also carry a small first aid kit under the seat of the truck and Suburban, containing bandages, antibiotic ointment, burn medication, sterile gauze, tweezers, aspirin, sterile eye wash, and cold tablets for ourselves and our eight-year-old son, David. In the glove box is a smaller snap-open plastic box with Bob's oral diabetes medication, my blood pressure pills, and a few aspirin. This has come in handy many times when someone forgot to take prescribed daily medication or a headache suddenly popped up. As the glove box does get hot during the summer, this small stash of meds is rotated routinely to make sure the strength does not fade.

Learning to use your kit

No matter how comprehensive your medical kit is it can be useless or even harmful if you do not know how to use it safely. You don't have to have extensive medical treatment to handle most emergencies that occur in real life. Most of ours consist of splinters, minor cuts and scrapes, sprains, and an occasional head cold or the flu. While these are scarcely life-threatening, they are uncomfortable and the afflicted party sure appreciates quick, competent aid.

Our family is lucky; I have spent a lifetime as a veterinary field technician riding on calls and acting as an assistant on everything from broken legs to pneumonia. Bob is a Certified Nurses Assistant (CNA) with additional military medical training in Vietnam.

But you'd be surprised at how much free medical training is out there for you to pick up. Many communities provide first-aid classes, including invaluable cardiopulmonary resuscitation (CPR) training.

During these classes, ask questions to boot up the amount of knowledge you receive. Attend volunteer firemen's training sessions, as available (again, ask around). Ask your veterinarian if you could accompany him/her on calls one or more days a week free in exchange for the knowledge you gain. Yep, I know, they're animals, not people, but basically, a mammal is a mammal, especially when it comes to shock, wounds, and common illnesses such as pneumonia.

Pick up a good first-aid manual (which should be in your large medical kit at all times) and a book or two from a preparedness company which details medical treatment when no doctor or dentist is available. Then *read* these manuals carefully. I know they're not great reading, but they can save someone's life. Share the reading with your spouse or older children, and even practice at home. It can be interesting, learning to suture gaping wounds on a piece of that chicken you're having for dinner. After it's butchered and ready to cook, of course. I'm not that morbid.

Real life medical treatment basics

While some survival first aid manuals assume your family's injuries will need treatment for nuclear blast and gaping wounds, in reality most will be of a much more mundane level no matter where you are, from arctic tundra to urban sprawl. They will consist of minor cuts, scrapes, slivers, blisters, the flu, colds, a fish hook in the skin, etc. We have lived for years in very remote locations and, although the worst injury any of us sustained was Bob's green stick fracture of his leg in a snowmobile accident, the most painful was my severely sprained ankle, suffered when I missed a step going downstairs in our farmhouse in "civilization."

Let's look at some real-life possibilities and what to do about them, assuming that there is no doctor or hospital available. Remember that if trained medical help is available, one should always consider this course first as many conditions can be made worse by incorrect diagnosis and treatment.

Hypothermia

Believe it or not, hypothermia (the condition where the body temperature is lowered below normal) kills more people in survival/stress situations than does gunfire, wild animal attacks, poisonous reptiles and spiders, wounds, or drowning.

Hypothermia has many causes, from shock following an accident to remaining outside in cold weather without adequate clothing or shelter to getting dunked in icy water—even for short periods of time. It is definitely something to watch for in any survival situation.

Identifying hypothermia can be a problem with the uninitiated, as it comes on slowly and the person still can walk and talk. But by paying careful attention, one can usually notice body shaking, paleness, and a tendency toward poor judgment and/or speech that doesn't make sense.

Taking the victim's temperature, you will quickly see that it is subnormal.

Hypothermia must be treated vigorously and immediately. Warmth is the key. As the body has lost its ability to warm itself, simply putting a blanket around the person is not enough. Build a warm fire. If the victim is wet, get them into warm dry clothes quickly. Warm a blanket or sleeping bag, then wrap it around them while they sit or lie in front of the fire. If they are not too bad, a drink of warm coffee or tea often helps. But do not give anything to eat or drink to a victim that is dazed or unconscious.

If nothing else is available, have one or more persons crawl into a sleeping bag or blankets to provide bodily warmth to the victim. Then keep the person warm and dry until they are fully recovered. You don't have to be a mountaineer to suffer hypothermia. I have had several encounters: falling through thin ice while crossing a beaver dam, getting stuck out in an unexpected blizzard in June, and getting drenched in the rain while making a mile and a half canoe portage in Minnesota's Boundary Waters. Hypothermia can be just plain uncomfortably miserable, but it can also kill.

SUGGESTIONS FOR YOUR MEDICAL KIT (LARGE)

ITEMS	USES
thermometer	detecting fever
aspirin/Tylenol	fever/pain
zinc lozenges	head off colds/flu
cough/throat lozenges	comfort with colds
antifungal medication	fungus infections of the skin
antibiotic ointment	heal cuts/abrasions
eye medication	infections/irritations
oral expectorant	clear lungs, reduce coughing
burn medication, such as *Burn Free*	reduce severity and pain from burns
oral electrolytes	treats dehydration
rolls of 2" sterile gauze	covers wounds, control bleeding
several packs 2" sterile pads	covers wounds, gauze controls bleedng, holds medication in place
rolls of elastic leg wrap	supports sprains, holds meds in place controls bleeding, protects legs
rolls of sterile cotton	cleaning area, controls bleeding, etc.
alcohol, soap, Betadine	cleaning, disinfection, wound healing
oral antidiarrheals	treating moderate diarrhea
any family daily meds	maintaining health
oral antibiotics/sulfas	treating systemic or local infection
injectable antibiotics/sulfas	treating systemic or local infection
injectable ephinephrine	shock, as in drug allergy
injectable antihistamine	allergies; bee sting
surgical instruments, such as forceps, needle holder, scalpel w/blades, scissors, etc.	facilitate minor surgery
assorted sizes suture material; absorbable	allows suturing of gaping wounds
stethoscope	monitoring vital signs
sterile needles and syringes	giving injections
sterile IV kit (if experienced)	makes IV injections possible for severe dehydration
IV electrolytes	severe dehydration, shock
first-aid manual	instructions

Of course your family's personal medical kits (small, medium, and large) will probably contain different items, depending on your foreseeable needs, medical experience, and preferences. And you will probably think of many more items that would be provident to carry, especially in your large kit. There are no hard and fast rules, only suggestions. The main thing is to be prepared—and confident.

Wounds

Most wounds that folks suffer in a survival situation are relatively minor, and though they may be uncomfortable and even bleeding, they are not usually life-threatening. The thing is not to panic. A little blood looks like a lot, especially when it is on yourself or a loved one.

If the wound is combined with possible other injuries, such as following a tumble down a rocky slope, you have to first assess the possible damage. Could there be a broken bone? A concussion? Internal injuries?

Don't panic. However, if you suspect such complications, do not move the injured party unless absolutely necessary, and then do it with great care.

Talk to the victim. He can usually tell you a lot about where he hurts and how much pain he is in. If the only injury seems to be the wound, reassure the victim and begin treatment.

Check the wound. Is it visibly dirty? Is the blood simply flowing from the wound or is it spurting? In survival situations, more people die from infected wounds than bleeding to death.

If the wound is relatively minor and the bleeding is minimal, you'll want to gently clean it before any attempt is made to bandage it. Nothing causes infection more than bandaging an unclean wound, even if it contains no visible dirt. Remember that deadly staph organisms are commonly found on human skin.

A good way to clean most wounds is to gently bathe the area with mild soap and water. Mop away from the wound, as one would sweep a floor, instead of scrubbing back and forth. The latter only moves bacteria around rather than removing it from the area. Rinse or soak the area well, removing any debris carefully with sterile tweezers.

When the area is clean, pat it dry with sterile gauze or air-dry it, then apply Betadine or antibiotic salve. We use Betadine for deeper wounds, and antibiotic salve for lesser injuries. Minor wounds seldom require bandaging, healing quicker by air exposure. Deeper wounds and ones in areas where they will be constantly irritated by clothing or work should be bandaged. A simple adhesive strip usually does the trick.

If the wound is bleeding quite a bit, simply applying pressure to the area with a sterile gauze pad will usually stop it within a few minutes. Where tourniquets were once advised, it has been found that more damage was done by the tourniquet than the bleeding would have caused in most instances. The application of firm pressure directly to the wound is very effective. After the severe bleeding has been stopped, gently clean the wound, but do not destroy the clot that has formed or bleeding will probably resume.

Should you be dealing with a more severe wound, covering it with a Betadine soaked (but not wet) sterile gauze, then a plain sterile gauze square, then adhesive tape is usually sufficient. If the edges of the wound gape or there is a flap of skin hanging down, either gently match the edges with butterfly adhesive strips or suture them, if you have the experience. Remember that most wounds will heal fine without suturing, especially with a little help from gentle butterfly adhesive strips. Sutures that are too tightly drawn will cause pain and scarring.

Never bandage a wound tightly with gauze bandage or anything else. This will restrict circulation and can cause pain and severe problems and even gangrene.

In the following days, keep the wound clean and dry. Change the dressing as needed, usually twice a day, leaving the dressing off and the area open to fresh air and sunlight as much as possible. This will greatly reduce the healing time and reduce chances of infection. Bacteria love damp, dark, warm areas, including a wound which is bandaged.

Watch for ugly redness or a fever in the patient, which would indicate infection in the wound. In this case, keep the area soaked in Betadine and give the patient antibiotics for 10 days, even if they seem better within a day or two. Immersing the infected wound in a hot Epsom salts solution also helps reduce pain and swelling along with cleansing the area.

Simple pain and swelling from the injury can be alleviated by plain aspirin, taken orally. Do not give aspirin immediately following an injury if there is a possibility of internal injuries, as aspirin may enhance hemorrhage. Do not give aspirin to young children. Use Ibuprofen instead.

Colds and flu

These common conditions are bad enough when things are fine, but are downright miserable in a survival situation. And remember that stress helps these overcome your body.

At the first sign of a cold or the flu, do those things your grandmother told you: keep warm and dry, rest, and drink plenty of fluids. Then add vitamin C and zinc lozenges, and most folks can overcome that mean cold or flu in a few days. If you need to alleviate symptoms, such as fever, runny nose, or coughing, take a cold/flu medication that covers your symptoms. By now, you probably know what works best for you and your children. The key is to have the medication on hand.

If the cold or flu lasts for longer than 10 days or seems to get worse, it may have turned into bronchitis or pneumonia, and antibiotics are necessary. Remember that home treatment is only for when no doctor or hospital is available.

Sprains

Believe it or not, sprains are one of the most common injuries in a survival situation. And often one of the most painful. The sprain can arise from walking over debris, logs, rocks, and even urban curbs. It can come from a fall or even an ankle turning over. (Your family will experience less sprains if they wear good, sturdy footwear, not flats or sandals. Ankle support is very important.)

When a sprain is new, immerse the affected part in cold water or apply ice packs to reduce pain and inflammation. I've found that when I take two plain aspirin immediately following such an injury that it greatly reduces both pain and inflammation later on.

If possible, rest the sprain, keeping it immobile and elevated for as long as reasonably possible. I continue taking the aspirin to keep down the inflammation. If you must move about, gently wrap the area with an elastic bandage to support it. Do not wrap area tightly or you will restrict circulation and make the pain much worse. Use a cane or crutches if the sprain is in a foot, ankle, or knee to reduce the amount of weight put on the injury. If the sprain is in the hand, wrist, elbow, or shoulder, keeping the arm in a sling will greatly reduce the pain and help it heal.

After a day, begin using hot Epsom salts soaks or packs to reduce the swelling and pain. And remember, the more you use a sprained joint, the longer it will take to heal and it may not ever heal completely if you persist ently use it before it heals. Rest is the key.

Slivers and spines

Getting a sliver or sticker of some kind is awfully common, especially in a survival situation, when one may be building a wood fire or foraging for food. Most of the time you can simply get hold of it and pull it out and be no worse for wear. But sometimes it is in too deep and painful and seemingly impossible to remove.

For relatively minor, but painful slivers, I use a sterile hypodermic needle, choosing the gauge (diameter/size) to fit the sliver size. Most smaller slivers are removed very easily with a 20-gauge needle. Now I use a hypodermic needle for several reasons, as opposed to using a sewing needle. First, and most important, they are hollow. This allows one to slip them into the sliver track with little pain, as less bulk is pressing on that tender skin. They are also sharp, which lets me carefully pick away the skin layer above the sliver which has no feeling because there are no nerves, until the sliver is exposed and can be either snagged with the needle and drawn out or picked up with a pair of sterile tweezers and removed.

With larger slivers, I use an 18-gauge needle, which does the same thing but is a bit stronger. When the sliver is very painful, using a local anesthetic, such as oral medication or antibiotic ointment containing an anesthetic on the area about 10 minutes before the procedure, helps a lot.

The main thing is to keep the sliver aligned with its track, and not to pry it upright in removal, which is extremely painful.

Once the sliver is out, a little alcohol or Betadine will disinfect the area and let it heal quickly.

I've discovered a great treatment for small stickers and cactus thorns which break off when you try to remove them from tender skin. Should you or a family member fall into a cactus or other plant with fine stickers, simply coat the area with *Shoe-Goo* or *Sportsman's-Goo*, which is a clear silicone-type product. Just a thin coat is fine. In about fifteen minutes it will be dry, and you can just peel it off, complete with all of the painful stickers.

Of course, there are many other possible injuries and illnesses. With a little advance preparation and study, you'll be surprised at what you can glide smoothly through. There is seldom any benefit to panic; a positive mental outlook can save lives. Δ

Chapter 10
Emergency gear for your vehicle

By Jackie Clay

Just as everyone knows it's sensible to carry an inflated spare tire, a jack, and lug wrench for your vehicle so it won't get stuck, carrying emergency gear for you and your family's safety and comfort makes just as much sense.

A good flashlight with fresh batteries is a must in every vehicle. We carry ours under the driver's seat and check it frequently to be sure the batteries are in good shape. Nothing is more miserable than having some sort of trouble in the night and being in the dark.

Carry as much emergency gear in your vehicle as you can conveniently. If you have the room, carry a gear box, such as we have under the shell of our pickup. It's large enough for a sleeping bag for each member of the family, a change of warm socks, a warm jacket, a small bow saw, a small propane stove and cartridge, candles, a few butane lighters, a pan, and other gear. Most vehicles will provide room for nearly this much emergency gear in a trunk or other little used cargo space.

Every vehicle should have at least one warm blanket in it. Even during the summer. At night, or during a rain, it can get cold without the heater to run periodically.

A few warm clothes, tucked in a dust-proof bag in the trunk, can be a lifesaver, especially in wet or winter weather. Be sure to have something for all members of the family.

Heat, in the form of a candle or propane stove or lantern, is a good idea. Be careful when using a candle, as they can easily be knocked over inside a vehicle, causing a flash fire of toxic fumes. But even a smallish candle can provide enough heat to keep a family from freezing to death inside a vehicle that is disabled. It will also provide light to attract the attention of rescuers and help prevent the vehicle from being struck by passing traffic or a snowplow.

When using any form of heat, whether it be from starting the vehicle from time to time to keep warm or a candle, be sure to crack open a downwind window, preventing carbon monoxide poisoning. This odorless toxic gas has killed hundreds of stranded motorists, silently, without warning.

A means to make a fire is a must. The fire can draw attention to you if you need help, it can cook food and can keep you warm. Fire starters can range from butane lighters to windproof matches to a flint, magnesium, and steel kit. This last is the best.

Always carry vital motor and radiator fluids, as well as a gallon jug of water in the trunk. These include anti-freeze, brake fluid, transmission fluid and, most important, motor oil. For the lack of a quart of oil, you could blow the engine. And it could be a long walk to "civilization," especially knowing you face a $1500 engine job.

Stick a good shovel in the trunk. If you don't have room for a full sized shovel, cut the handle off one to fit, leaving as much of the handle as you can, or use a folding Army shovel. The new Army issue folds down to 12 inches. A shovel could save your butt if you get stuck in snow or mud.

Likewise, always carry a set of jumper cables. You can usually find someone willing to give your dead battery a jump, but he doesn't always have his own cables.

A basic tool box is essential. Simple tools, such as a couple of screwdrivers, an adjustable wrench, a pair of channel lock pliers, wire cutters, a battery terminal cleaner, hammer, electrician's tape, and duct tape can save the day. Even if you don't know much about auto mechanics, a knowledgeable passer-by may be willing to help, if he has the tools.

In this vein, be sure you have a box of assorted fuses tucked away in the glove box. Something as simple as a blown fuse may disable a vehicle. (One tip here, if you are lacking a correct sized fuse, rob a like sized fuse from something non-essential, such as the radio, replacing it when you get to a service station.)

Have drinking water with you at all times. This can be that gallon of water in the trunk, but if it is, change it from time to time, keeping it fresh. People have died from lack of water, and not always in southwestern deserts.

A fire extinguisher is a good addition to a vehicle's emergency gear. I've had a van I was driving catch fire without warning. Luckily a fire extinguisher was handy. All that burned up was the wiring under the hood. Check the fire extinguisher at least twice a year, to make sure it is charged.

Carrying a good, moderate sized first-aid kit is essential. Many of our little emergencies happen periodically. The kids skin a knee, your husband gets a steel sliver in his hand, you burn your arm checking the radiator. Nothing earth-shattering, but, at the time, emergencies nevertheless.

Find a sturdy, flat plastic box that will fit under the seat and fill it with commonsense first-aid gear: assorted adhesive strips (cuts, scrapes, blisters, etc.), eye drops, a needle and tweezers to remove slivers, antibiotic ointment,

Betadine or iodine, a roll of two-inch gauze and gauze squares (sterile), burn ointment, aspirin, Tylenol, and anti-diarrheal tablets. Tailor your kit to your family's lifestyle. Make it comprehensive, but not bulky.

A little "survival" food, tucked in a bag or box in the trunk or under the seat, is a great idea. We certainly won't starve without a few meals, but I, for one, would rather have a little something to munch on than to go without. This is especially true during times of duress. This survival food can be high energy bars; small cracker and cheese or peanut butter packages; pop-top foods such as chili, stew, wieners, MREs (military meals ready to eat), dehydrated fruit slices, etc. The main thing is to choose foods that are not affected by heat/cold, will not go stale quickly, and will provide energy and satisfaction.

Now don't laugh, but toilet paper should be in every vehicle. Yep, you can do without it, but wiping on sticks and rocks is one aspect of my Native American heritage that I don't want to return to. Besides, toilet paper can be used to start a campfire, wipe the oil stick when checking the oil, stop minor bleeding, clean up messy kids, and blow your nose.

A transistor radio, especially one of those little ones that is both solar-powered and rechargeable with a built-in crank, would be handy to keep track of weather and/or riots. And don't forget snow chains for winter.

I'm sure, with these hints, you can think of things you'd like to include in your family's personal vehicle emergency kit. No two families are the same in their needs, but the point is to be ready for the unexpected, then relax and enjoy life. Δ

ON YOUR PERSON

On your person you should carry some or all of the following:

- means to make a fire, whether it's a butane lighter, wind-proof matches, or better yet, one of those pocket-sized magnesium and flint and steel tools such as you find sold at the Preparedness Expos.
- pocket knife or *Leatherman* type tool with can opener, pliers, screwdrivers, saw, tweezers, etc.
- vital medical prescriptions
- emergency money

Also wear proper footwear, not sandals. If you wear a fanny pack you can carry more, such as a tiny fishing kit, consisting of hooks, line, and sinker placed in a film canister. Also some granola bars, a garbage bag (raincoat), and a tiny flashlight. Some people even carry a gun in their fanny pack.

Chapter 11
Rifles, shotguns, and handguns for survival

By Massad Ayoob

You bought it for a worst case scenario. If that doesn't come to pass, you'll still own it. I'm talking about rifles, shotguns, and handguns that'll get you through bad times and good. Consider this approach: Prepare yourself with things you can use if society doesn't break down.

Experienced rural people have known for a very long time that having a generator just makes good sense. I'd be surprised if there's a hospital in the land that isn't outfitted with generators in case the power goes out for some reason. If you have room to store it, isn't it just practical to buy staple foods, toilet paper, sanitary napkins, et al in bulk and stockpile them, saving money compared to the usual urban "wait 'til we run out and then buy as needed" paradigm?

So it is with the guns. Let's look at some rifles that would be awfully good to have on hand if society breaks down and would be good to have on hand anyway.

.308 rifles

Springfield M1A. As currently produced, this is a splendidly-made example of the M-14 rifle, a design that goes back roughly half a century and is essentially an improved WW II M-1 Garand. With up to 20-round magazines still readily and legally available, it fires the 7.62mm NATO/.308 Winchester cartridge. Ammo is abundant. The rifle is utterly reliable.

The M-14 was constructed for a projected battlefield in which it was thought that the hedgerow fighting and house to house combat of the European Theater would be the small arms battle plan of the future. This is remarkably close to defending a home against armed invaders on foot or in cars and trucks in a "post-Armageddon" scenario.

No Armageddon? No problem! A five-round magazine makes this accurate, reliable autoloader an excellent hunting rifle in rugged country. More than a decade ago, in Namibia, I used a sniper grade M1A Springfield to shoot a springbok at 350 yards. The local rules said "no semiautomatics," so I used the gun without a magazine as a single shot with just the one round in the chamber. Ben Mozrall, then head of the New Hampshire State Police SWAT team, shot a 1.5-inch group with this same rifle at 200 yards. In a weak moment, I sold it with its exquisite Leupold scope to John Groom. Later I came to my senses and bought a stainless Match Grade Springfield Armory M1A to replace it. Target shooters, be advised that this is the rifle that kicks butt in national class High Power competition. The prices are excellent these days. An added feel-good bonus: no American gun manufacturer has been more vocal about individual citizens' Second Amendment rights than Springfield Armory.

Other fine .308 "battle rifles" include the European **FN SLR** and **HK91**. In the score of years I've owned it, I can proudly say my Heckler and Koch HK91 has never malfunctioned.

.223 rifles

With ammo readily available, the 5.56mm NATO/.223 Remington round will get you through the darkest of human conflict nights. It's the choice of not only the entire US military, but most of the cops. If I was going into battle, I'd want the superb ergonomics (speed reloading, ambidextrous use around cover in any configuration, fine balance) of the **AR-15**. For my own needs, the short, handy, super accurate **Steyr AUG** is a personal favorite, but I bought mine back when the price was reasonable. For a "best buy" today that's still user-friendly, you can't beat the **Ruger Mini-14**, which, in stainless steel, is the mainstay of this magazine's publisher, Dave Duffy, and its senior editor, John Silveira.

The .223 is also ideal for "back porch to garden and beyond" sniping of local animals who figured you planted your garden for them instead of you. With light, high velocity hollowpoints, such a rifle is still a devastating manstopper for home defense, offering light recoil with high hit potential on multiple targets at close range under stress in short time frames. And .223s are always fun for recreational plinking, which is what my youngest has used hers for (**Olympic Arms AR-15** with C-More red dot electronic sight) since she was twelve.

7.62 X 39 auto rifles

This ComBloc cartridge, Kalashnikov's updating of the first (German, WW II) assault rifles, approaches the classic American .30/30 hunting round in effectiveness when used with a soft-point hunting bullet. There are essentially three popular choices. A semiautomatic version of Kalashnikov's own **AK-47** will indeed work through mud or blood or crap or flood, but it won't be terribly accurate. Good for close fighting, not great for recreational use at significant distance. Ruger's Mini-14 in this caliber is known as the **Mini-30**, and with hunting-legal five-round magazines standard (and longer magazines available "aftermarket") makes more sense as "survival situation insurance for now, and a lightweight, light-kicking deer rifle for later." Finally, there's my own choice in this caliber, the **SKS**. This old ComBloc beast is at least twice as accurate as the AK-47, rivaling the Mini and the AR. It's as ruggedly reliable as the AK, generally has a better trigger, and 10 shots instead of a bigger magazine is its only shortcoming. They're cheaper than dirt, as gun prices go, even now. I prefer the well-made commercial Norinco guns from China, and the Russian Paratrooper versions, because they both have short stocks that work well for petite females, or for adult men in heavy cold-weather coats. The same softnose hunting ammo you'd use in a Mini-30 can make the SKS a useful "farm gun" once any perceived societal crisis has passed.

I don't care for AK rifles much. Factory magazines of 20- and 30-round capacity for the Mini-30 are limited to cops under current law. For an ordinary law-abiding citizen getting started, I'm not sure you can beat the cost effectiveness of the SKS. I won one in a match, gave it to my oldest kid, and then bought an identical SKS for my youngest. In the best of all worlds, I'll have died of old age and each of my daughters will have an SKS, ammo, and accessories secured safely in their homes for their kids, far in the future.

.22 rifles

Survivalist guru Mel Tappan said you should have it. Gun experts say you should have it to teach marksmanship to your kids and other new shooters. Fast and in-the-eye-socket accurate, the gun we're talking about is a top-quality autoloading .22 rifle. The ammo is cheap, and so light it's portable. A box of 50 rounds of .22 Long Rifle is the size of a small survival ration of wooden matches.

For accuracy, reliability, and cost effectiveness, you can't beat the time-proven **Ruger 10/22**. I hate to sound like a PR man for Ruger, but that's just how it is. There are more aftermarket accessories (folding stocks, match-grade barrels, extended quick-change magazines, etc) available for the 10/22 than for any other .22 rifle. A close second on my list would be the **Marlin Model 77**. Over a million of these neat little rifles have been made. They're accurate, they're reliable, and the only reason I don't rate them over the Ruger is that you can't readily get extended quick-change magazines for all of them. (Most Marlins have tubular magazines, which are comparatively slow to reload.)

I enjoyed my 10/22s and my Marlin 77 before any threat to society appeared. I'll enjoy them once it's over. An accurate .22 caliber semiautomatic rifle will still pick the squirrels off your bird-feeder, still zap the woodchuck in your backyard garden from your kitchen window, and still put a 50-grain bullet through the rapist's eye and into his brain as he breaks through the bedroom door.

Other rifles

"Other rifles" is a world to cover, and I want to keep this short. Sure, a lever action .30/30 will get you through the night, but it's slow to reload in an emergency involving high-volume shooting. Yeah, the Lee Enfield .303 Jungle Carbine was the best bolt-action battle rifle of its time and not a bad deer rifle even today. Yes, the Steyr-Mannlicer rendition of Jeff Cooper's Scout Rifle is superbly crafted. Liked mine, sold it anyway. These are bolt-action rifles. When you show me how I can get off five shots in one second with a bolt gun at "down the hall distance" like I can with a semiautomatic, enlighten me, OK? Until then, a gun that would do to protect one's innocent family against many-to-one lethal odds needs to be semiautomatic as far as I'm concerned, and for purposes of this article, the topic is worst case scenario family defense guns that can be useful later. If the emphasis seems to be on later, I apologize, but we can't ignore realistically-perceived short term needs now.

Shotguns

The shotgun is a versatile firearm. But is it the ideal defense gun for a worst case scenario…or even for the best of times?

Los Angeles cops used to call it "the tube." Westerners called it "the scattergun." A whole lot of folks have called it the single most versatile of all hand-held firearms, and within its limitations, they may be right.

The "shotgun" is so named because when you drop its hammer on the ammo it was primarily designed for, it expels a cloud of pellets known as "shot." There is birdshot, a great number of tiny pellets, suitable for killing light-bodied creatures like the fowl of the air. There is buckshot, fewer pellets but each larger and heavier, which at close range act like so many old-fashioned, non-expanding pistol bullets striking all at once. These are designed to kill deer at short ranges, and other animals in the 130 to 300 pound range, animals like…oh, Hell, you figure it out. Finally, there are shotgun slugs, which turn what was once known as a "fowling piece" into .70-plus caliber rapid firing heavy rifles in close quarters.

You use fine birdshot to shoot "upland birds." You shoot heavier, coarser birdshot for "waterfowl," ducks and geese that you'll have to shoot at longer range through heavier feathers and body mass if they are to come down from their migration routes to grace your table. Buckshot ("double-ought," as it's known historically) sends nine .33 caliber pellets from the muzzle of a 12-gauge shotgun at once. It's roughly the same as shooting something nine times in the same millisecond with a .32 caliber automatic pistol and old-fashioned ammunition. Cops use it when they anticipate close-range gun-fights, and African hunters used 00 Buck when going in close on wounded lions and leopards in the thickets. Finally, the 12-gauge rifled slug—an ounce of lead almost ¾-inch in diameter, flying at 1100 to 1600 feet per second—is not only the close range "shotgun deer load of choice," but is what some professional hunters in Alaska and elsewhere resort to when they have to go into the underbrush after wounded grizzly bears. It is also the ammunition the NYPD Stakeout Squad switched to after a few shootings where even the deadly "double-ought buckshot" didn't put their armed opponents down fast enough.

Good news with shotguns

Obviously, the shotgun is versatile. So are its design options. You can get a single barrel (too slow for anything more serious than sport, but cheap). You can get a double barrel, anywhere from $200 for a second-hand Stevens 311 that will get you through the night, to six figures (yes, you read it right, over $100,000) for a super-grade shotgun custom made in Europe. Or, you can get a functional pump-action shotgun (Mossberg or Remington Express) for around $300 or less. A semiautomatic shotgun will start a bit higher than a pump.

Bad news with shotguns

When the strength is versatility, the weakness will generally be found in specialty. The history of the human experience seems to be that whatever works half-ass well for everything will be found specifically deficient in something. So it is with shotguns.

So, what we're talking about here are shotguns that might be bought by someone anticipating a worst-case scenario after a societal breakdown. One study of LAPD use of (12-gauge pump) shotguns in actual gunfights indicated a 58% hit potential. That means out of every two shots fired in anger, slightly more than one was likely to put one or more 00 buckshot pellets into the opponent. If we translate that to a multiple-opponent gunfight scenario, it tells us right up front that a one-or two- or even three-shot gun will give us a limited future against multiple opponents.

This tells us that for defensive purposes, we need a baseline of a "magazine-fed shotgun," which will hold an absolute minimum of four shotgun shells in its reservoir, and if time allows the defender to load the gun all the way up, with one more shell in the firing chamber, a minimum of five shells capacity. **Important note:** There is **no** magazine-fed shotgun made today to this writer's knowledge which incorporates a firing pin safety. This means that **any** shotgun kept for emergency defense purposes needs to be stored with its firing chamber empty, so that it cannot accidentally discharge if dropped or struck on either end.

What not to look for

Don't tell me that your shotgun is cool for every distance because it has interchangeable chokes. You won't have time to change the chokes in the moment it takes you to transition from your first opponent at your doorway to the one 40 yards away in the parking lot who is aiming his stolen rifle at you.

Don't tell me that your shotgun is emergency versatile as opposed to situationally versatile because it can shoot birdshot, and buckshot, and slugs and all the rest. The versatility requires time for you to load the right shells into the magazine! That time may present itself to a police SWAT team that has the suspect in the crosshairs of multiple gunsights. It won't be there when you have to quickly change gears between perpetrator A on the front doorstep and perpetrator B, also armed with a rifle, way out there at the barn.

Don't tell me you load your shotgun with birdshot for home defense because, "It will blow away a bad guy, but it won't shoot through the walls and hurt the kids." A whole bunch of us have put our fists through the kind of sheetrock walls that separate our bedrooms, but none of us was ever able to put our fists through the opponents we punched in the stomach, as hard as we tried. If it will go into your opponent enough to blow away his internal organs, it will go through the wall and endanger your children too. If it won't go through the sheetrock your fist can go through, don't count on it to keep a homicidal antagonist you hit dead-center with it from killing your babies.

I'm sorry if that takes away something you were counting on, but somebody had to tell you in time.

Recommended shotguns

Your defensive shotgun should be a pump action or a semiautomatic. For professionals who keep their guns clean—and know how to break them down to do that—I strongly recommend the semiautomatic. For people who tend to neglect their equipment, I would most strongly recommend the pump gun.

There are damn few semiautomatic shotguns that will cycle 100% with light loads, medium loads, full power loads, and Magnum loads, no matter what you may have read in gun magazines. I have been to "gun writer seminars" where the hosting gun companies fed us up to the gills with shrimp and booze and paid for our nice hotel rooms, and some of my colleagues wrote great things about the guns that malfunctioned when we all tested them. I wrote the truth. That's why I'm not invited to some of those seminars anymore, but my duty is to you, not to the gunmakers.

When in doubt, buy a **Remington Model 870 Express** or a **Mossberg 590** pump gun as first choice, a **Mossberg Model 500** as close second choice. If you want a 12-gauge autoloader, buy a **Remington Model 11-87** or a **Benelli Super-90** or a **Mossberg Jungle Gun**. I say this not as the gunwriter who supped at the tables of the manufacturers—actually, I am that, I'm just not speaking as such.

I speak now as the full-time firearms instructor who sees what works and what doesn't, 20 to 40 students per combat shotgun training session who hose buckets of buckshot downrange at high speed and find weak points in their guns that people who shoot light trap and skeet loads never discover.

Any of the above-recommended guns will serve you well in an emergency. In the lighter-kicking 20 gauge, spewing 20 .25 caliber buckshot pellets per pull of the trigger, I'd recommend the **Remington semiautomatic Model 1100**, especially the lightweight LT-20 model with short stock that fits smaller people better and causes no shortcomings to those with taller stature or longer arms.

Handguns

Let's look at handguns for emergencies. A single gun has to be multi-purpose:

- Accurate enough to kill an animal some distance away for food.
- Powerful enough to stop a threat coming in on you quickly.
- Holding enough ammunition to allow you to shoot fast and straight and, if necessary, reload quickly and shoot fast and straight some more.
- And, when you don't worry about "panic in the streets" any longer, a gun that will still serve you well for recreational shooting, home and personal

defense, and for the chores that firearms have historically performed at rural homesteads.

Single action frontier style revolvers and single shot pistols and derringers are out. Not enough firepower.

22s have a place in a survival kit for shooting small game for subsistence, and the ammo is certainly cheap and easy to carry in volume. You have to determine what you're most likely to need to shoot. .22s are cool for squirrels. In the anti-personnel function, however, they have proven effective for only two classes of people: highly trained Israeli intelligence operatives and Mafia assassins. Both seem to have the knack of getting behind the opponent and shooting him through the base of the skull. If this doesn't seem a likely scenario for your needs, then as useful as a low-powered .22 handgun is around the rural home, you can do better for an all-around handgun whose purposes include emergency defense.

.38 Special/.357 magnum revolvers

The .38 revolver won't take the longer, more powerful Magnum cartridge of the same diameter, but a .357 Magnum revolver will fire shorter .38 Special cartridges. Unless there's a compellingly necessary cost saving, in guns of the same size buy the .357 and you'll have a better chance of resupply if the future is such that ammo becomes scarce. The small ones shoot only five rounds, a few of the new bigger models hold seven or eight, but for the most part a .38 or .357 revolver will be a six-shooter. Have some speed-loaders handy, of the proper size for that particular revolver, and you can learn to quickly reload six shots at once in less than six seconds.

These guns are sturdy and reliable. If all the gunsmiths were going to disappear, I'd choose the built-to-last-forever **Ruger**. The **Colt**, **Smith & Wesson**, and **Taurus** brands are all good choices, too.

.45 semiautomatic

Survivalist guru Mel Tappan recommended this gun, specifically the **Colt Government Model** or one of the other 1911A1 variations, as that pistol was known when it served the US military for most of this century. All good-quality .45 ammo is powerful enough to trust in a fight, though the hollow-points work a lot better and are less likely to shoot through the bad guy and hit a good guy. Many of our readers were trained on this gun in their military past. The gun you are most trained with is the one you'll fight best with, and when you are judged later for your actions, being trained with the gun you used is a plus. This is an advantage for the service revolver mentioned above if you were trained with one in a former life as a cop or security professional.

If you like the .45's power but don't like the necessary "cocked and locked" carry mode for the 1911 gun, you can get any of several modern double action .45 automatics. These require a long, heavy, intentional pull of the trigger to fire. **S&W**, **SIG**, and **Ruger** are all good. The **Glock** is another very reliable modern .45 automatic. In fact, it's been my experience that the .45 caliber versions are the most accurate pistols Glock makes.

9mm semiautomatic

Tappan didn't like this caliber, but that's because in his time most of the 9mm ammo available was impotent in terms of quickly stopping aggressive threats. Things are different today. You can get loads like the Triton Hi-Vel I carry in my own 9mms. A standard 115-grain hollowpoint will leave a 9mm's muzzle at 1100-something feet per second. The Illinois State Police "hot load" (so hot they won't sell it to the general public) sends the same bullet out of their **Smith & Wesson 9mm** autos at 1300 feet per second. The Triton Hi-Vel is rated at 1325 feet per second from short barrel pistols, and I recently chronographed it at more than 1450 feet per second out of the almost 5-inch barrel of one of my Berettas. This is ".357 Magnum country" in terms of potency. A pistol with 16 rounds of this 9mm, weighing the same as an eight-shot .45, is no contest for me: I'd go with the hot-loaded 9mm every time. (If you can't find Triton locally, we stock it at my place. Call toll-free (800) 624-9049 for catalog and ordering info. You'll need to mail or fax an ID card proving that you're at least 21.)

Alas, the 9mm choice presumes always being able to get the best ammo. A 9mm is very "cartridge dependent:" standard full metal jacket ("ball") ammo in that caliber is impotent for fighting determined opponents. If you may have to get ammo "wherever," go .45. If you have enough ammo stockpiled for your needs, hot 9mm has an edge.

Many readers have been trained with the Beretta M9, known in commercial circles as the **Beretta Model 92**, which the US military adopted over a decade ago. Again, thc gun you're trained and proven with is the gun to have. I know one guy who taught at Fort Benning whose "preparedness kit" includes two Beretta 9mms, one for each hip, in the same kind of holsters whose use he taught to soldiers.

The Beretta is a fine 9mm fighting pistol, but the **Glock**, the **SIG**, the Smith & Wesson, the old **Browning** and many others would also deliver noble service. With 9mm, remember, the trick is having enough of the right ammo on hand.

Points to consider. You'll hear or read people saying that you should have the Beretta 9mm because you can interchange ammo and magazines with military personnel. The same argument used to come forth for having the 1911 .45. Excuse me, but I think someone has seen one too many reruns of "Night of the Living Dead." If you really think cops and soldiers are going to share their ammo with private citizens they don't know during a crisis, I would like to talk to you about buying some oceanfront property in Kansas City. The Beretta 9mm and Colt .45 are fine guns, but buy them for good reasons, not stupid ones. The same is true insofar as the thought that you should have a .38 or .357 like the cops had before, or a .40 caliber pistol like so many cops have now. The police aren't likely to share with you either.

Training

Don't just buy the handgun, train with it! The handgun is the most difficult of firearms to learn to shoot quickly and accurately. Scumbag criminals in gunfights (as opposed to murderers standing on top of their helpless victims) only hit the good guys they shoot at with 15% of their shots or less. Cops minimally trained in marksmanship will go 17% to 35% in hit potential. The cops on well-trained departments—a skill level I truly believe you can reach in home study—will go to 40% or better hit ratio per shot in real-world gunfights. The highly trained SWAT officers and graduates of private academies approach or exceed 80% hit potential with handguns in actual shootouts.

So get training and practice.

(Massad Ayoob has spent slightly more than half his life, 27 years, as a sworn police officer. He has served for the last 11 years as a captain. He also owns the "Police Bookshelf," a source of gun literature and gun supplies, and the "Lethal Force Insitutute," which provides firearms training to both professionals and private citizens. Both can be found at www.ayoob.com.) Δ

Chapter 12
A survival garden

By Jackie Clay

Many of us have a garden and enjoy fresh vegetables during the summer and fall. Maybe we even have a few chickens for eggs and meat. But some of us want to extend our preparedness to what I call "hard-core" homesteading. This is serious homesteading, aimed at being able to provide your family with nearly all of its basic needs.

Luckily, most of us with a piece of out-of-the-way land can become nearly "store-bought-free," raising much of what we need in nearly the same way as did our ancestors.

There is a vast difference between this type of survival homesteading and stars-in-the-eyes, back-to-nature, recreational homesteading to relieve stress and provide enjoyment. The difference is not so much in how-to, but in discipline and learning.

The survival garden

It has been said that one can raise enough food for a family of four in a 50- by 50-foot space. While such an area can provide a goodly amount of food, there is no way a family could survive, year-round, off such a small patch. In reality, all that this is is a "house garden" for providing fresh produce such as greens, broccoli, cabbage, peppers, herbs, etc.

When one needs a garden to put up food, not only for the winter but possibly for a year or two, we're talking about at least an acre of intense cropping.

This includes a patch of wheat for grinding into cereal and flour, flour corn for hominy and corn meal, sweet corn for eating, canning, and dehydrating, and rows of dry beans as well as fresh beans (yellow wax, green, pole, etc.) for putting up. Here we stumble on the weak link in most folks' gardens. They say "We only use a few pounds of corn meal or dry beans a year," and they feel confident they can get by with just a few packages of such items, bought at the grocers.

But having lived in a wild corner of Montana, well above the "grocery line" (because of road accessibility), I can tell you that you will use many more pounds of these staples when you cannot eat from the store shelves.

And if there are no store shelves to choose from, we will all need to take care of our own needs at home. Remember, it takes more than one year to get a garden into full production. You can't just plow up a plot and expect to survive off of it, especially if you lack experience.

You can't grow everything, everywhere. Look at your local production capabilities. In New Mexico I could grow anything, but in the high country of Montana, nearly everything was a challenge even though I've gardened all my life. But we could survive from my Montana garden with potatoes, wheat, and beans along with a number of cold-loving crops we grew. What you need to do is put your energy into growing what will make a crop in your location.

Don't be afraid to experiment. Everywhere I've gardened I've grown crops that locals said "wouldn't grow."

To better use space, consider inter-planting as much as possible. Grow cornfield beans among the flour corn, radishes in the same row as carrots, peppers between rows of tomatoes (which act as windbreaks), pumpkins and squash next to a corn field where they can run into the corn after cultivation has stopped. (Don't do this with sweet corn or you will have a devil of a time picking the corn stumbling among rampant squash vines.) Inter-planting will do much to save garden space, a large consideration in survival gardening, especially when you must cultivate and till by hand.

Crops for a survival garden

Everyone who gardens grows some things just because they enjoy the taste. This is great, and we all do it. But in hard-core homesteading, we must consider our basic needs, as well.

We need to grow enough grain and corn for ourselves and livestock. This can be done by hand, in a relatively small plot, provided that our poultry and livestock needs are small. If you need more grain, say for cattle or horses, consider small scale farming with horses. This is a sustainable way of living as horses are easy to work, versatile, and provide manure for the fields. They also require no fuel to run. One team of moderate-sized horses can do as much work as a small tractor and cost little to maintain.

As little as an acre of ground can supply modest grain needs for a family homestead. Include a bit of rye, oats, and barley for variation. (There is a naked-seeded oat that is great for homesteaders, as at home one has a difficult task in hulling oats for oatmeal.)

Besides small grains, include your rows of flour corn for corn meal and hominy, being sure to include enough for livestock feeding.

Most folks have to double or even triple the amount of usual garden produce to allow for putting up as much each year as possible. Be sure to allow for lots of tomatoes for tomato sauces, and enough root crops, such as turnips, potatoes and carrots. (You'll eat a lot more "homegrown" when you can't run to the store for "quick" meals.)

With all survival garden vegetables, a family should buy only open pollinated varieties. This will enable folks to save seeds from year to year, which is not recommended with hybrids. Hybrid seed, while usually fertile, can not be depended upon to reproduce truly. And, contrary to popular belief, most of those old open pollinated varieties are good tasting and hardy.

Suggestions for a survival garden

- green beans, pole & bush
- sweet corn (various maturing dates)
- yellow wax beans
- carrots
- dry beans (several types)
- tomatoes (several varieties)
- potatoes
- turnips
- rutabagas
- cabbage
- broccoli
- cauliflower
- cucumbers
- onions
- greens of several types
- spinach
- lettuce
- peas (dry & green)
- pumpkins
- summer squash
- winter squash
- muskmelon
- watermelon

Perennials for the survival garden

Along with the vegetables, a hard-core survivalist should establish a good variety of perennial edibles. These include asparagus, Jerusalem artichokes, horseradish, garlic, perennial onions, and herbs for both culinary and medicinal use. Remember to encourage native perennial edibles which do well in your area. These may include prickly pear cactus (the fruits and pads are eaten as a vegetable), wild rice, wild greens, cattails, mushrooms, etc. In a survival situation, one truly appreciates variety in the diet.

The perennials have the advantage of having to be planted only once and usually expand on their own with little human help. And, like the annuals, which must be planted each year, a family can gather and put up many jars full of winter eating. I can wild and domestic asparagus, wild mushrooms, wild greens, cactus pads (known as nopalitos in the southwest), and dry many other wild and domestic perennials.

Small fruits are nearly essential

Nearly everyone has room to plant a good selection of small fruits. These include strawberries, raspberries, blackberries, rhubarb, blueberries, and so forth. Luckily, once a patch of each has been established, one can readily take divisions or replant sprouts to greatly increase their food-producing capabilities.

As with the vegetable garden, one should grow as great a variety of small fruits as possible, and enough of each to put up significant jam, preserves, and canned and dried fruit. In hard times, a good loaf of hot whole wheat bread spread thickly with homemade strawberry jam, or a steaming blueberry pie, makes the term "survival" a joke. We call it living good.

You quickly discover that small fruits are a wonderful treat that can be easily turned into strawberry shortcake, blueberry pancakes, rhubarb tarts, blackberry cobbler, etc. In hard times, you don't eat many candy bars; instead you substitute healthier fruit snacks and desserts.

Even picky eaters greatly enjoy dried fruits and fruit leathers which are easy to make at home.

Every homestead should include a small orchard

Even the smallest homestead has room for fruit trees. With the variety of tree sizes and shapes, you can choose full-sized trees which are tremendous producers, but take room and several years to begin bearing fruit. Semi-dwarf trees, which usually require only a 10- by 10-foot spacing, produce full sized fruit in moderate amounts and only take a couple of years to bear. Dwarf and "pole" trees, which produce full sized fruit in small amounts, can be raised on a patio in a portable tub.

A hard-core homesteader can get by with two each of several varieties to provide variety and cross-pollination. I'd suggest apple, pear, pie and sweet cherry, apricot, and plum for most gardeners. Of course, if you can grow citrus in your zone, go for it. We live in zone 5 and have two Brown Turkey Figs in a protected corner of our east flower bed—protected by the house from the killer winter north winds.

Now a lot of folks say they'd need acres and acres to reach this level of self-reliance in the food department. Not so. My grandmother did it on two city lots in Detroit. Instead of normal landscaping, nearly everything she grew produced edible fruit: peach, grapes, brambles, quince, asparagus, apple, crab apple, strawberry, etc.) Having gone through the Depression as a widow with two young boys to raise, Grandma knew how to fend off hard times.

But what about meat?

Like produce and fruit, a family can grow all of their meat requirements, right at home. Now few people actually like to kill to eat, but when it comes down to eating or not eating meat, most of us can find a way around our revulsions. After all, someone had to kill that steer that went into your Big Mac. It gets ridiculous when visitors won't eat a home-butchered beef roast but will buy a tainted, chemical-laden piece of plastic-wrapped roast at the supermarket and eat it with abandon.

Folks on a very small acreage will usually have to limit their meat production to poultry, rabbits, and perhaps a little goat meat. A small flock of chickens for egg and meat production, with a couple of hutches of rabbits and the castrated male offspring of the family dairy goats will do much to help out at the dinner table. Of course, a family with these reduced production capabilities will not eat meat every day, but it will be able to enjoy regular meals with meat as a feature.

The benefit of having only a small poultry flock, a few hutches of rabbits, and very few goats is that the feed requirements and labor requirements are also minimal. In such cases, a family can easily hand-raise and harvest all feeds necessary to maintain their meat and egg supply.

Small-holders can help supplement their meat needs by hunting and fishing. But remember, if times are hard nationwide, subsistence hunting will become very difficult in most areas, as it did during homestead days and during the Depression. The game quickly disappears with overhunting. Fishing holds up much better, so it benefits a family if they hone fishing skills before they are truly needed. (Besides, it's enjoyable family "work," as well.) For lucky backwoods dwellers who live near the seacoast or a salmon stream, fishing can well be the major source of family meat.

Folks with more acreage are in better shape to truly be meat self-reliant. Using horsepower to till moderate amounts of land, a family can raise enough small grain, field corn, and forage (hay and pasture) to maintain not only the horses but a couple of dairy cows or several dairy goats. Let me stop right here and address you folks who are saying, "Goats! No way am I going to raise those stinking tin can eaters!".

Goats do not eat tin cans, nor do they run around butting people, any more than do cattle. Goats are exceptionally clean, picky eaters, refusing to take a bite of the apple you just took a bite out of, and they'll dehydrate before they will drink from a bucket containing even one berry of manure. Only bucks in rut have any odor. While in rut they will spray their neck, belly, and chin with urine as an attractant to does in heat. So do elk and deer. The normal scent glands on a buck's head, which produce scent during rut, can be removed by surgery when the buck is an adult or during disbudding, leaving a scent-free male totally capable of breeding. Does never have an objectionable odor, and with neat droppings the pen is quite clean and odor-free with even minimal daily maintainance.

We've had both dairy goats and cattle, and we know the benefits and drawbacks of each. Both produce milk which is equally good-tasting. A goat often produces multiple offspring while a cow produces one calf a year.

Cattle are easier to fence in, but goats will do great in a pasture grown over to willows and brush as they are by nature browsers like deer. And, like deer, they can hop a four-foot field fence to enter your young orchard and strip the tender trees of their bark and twigs. Cows produce beef; goats produce chevron. Both are good, but different.

Chevron comes in carcass weights from between 20 and 100 pounds of dressed weight, depending on age. They are easier to cut up and handle, but their small carcass lasts a much shorter period of time than a 600-pound Angus carcass.

Remember that worldwide there are thousands more goats used for meat and dairy production than there are cattle. There are reasons, and economy is at the top followed by the quickness of meat consumption in areas without refrigeration. A 600-pound cattle carcass is likely to spoil before it's completely consumed.

Pigs are another cog in the serious homesteader's wheel of self-reliance. Not only can a few pigs easily be raised for butchering—being fed from home-produced feed, kitchen and dairy waste (skim milk is an excellent food), along with weeds, pasture, and hog-foraged feed—but they provide excellent meat with a carcass that is quite easily handled by the family. The bonus of hogs is that they produce lard, the only homegrown cooking fat easily obtainable.

Yes, I know about high cholesterol, but let me tell you that when you are working hard everyday to put food and other necessities on the table, your cholesterol will balance easier than your finances.

Homegrown dairy products

Okay, so far you have a good vegetable plot, small fruits, small grains, and an orchard and meat/egg supply started. It's time to think about dairy products, particularly milk and cheese. After that stored dry milk is gone, your family will want something to replace it. And what is more natural than learning to run a tiny kitchen dairy and cheese plant? All dairy products are quite easy to produce at home, and as with almost everything else, it's much better when homemade.

I've made cottage cheese, cream cheese, mozzarella, colby, cheddar cheese, sour cream, cheese spreads, balls, logs and sandwich loaves, ice cream, ice milk, sherbet, and more regularly at home, both from cow and goat milk. Butter and whipped cream are easier to do from cow milk, as the cream quickly separates out, floating to the top. Goat milk is naturally homogenized and it takes more "doing" to access the cream. Both animals' milk produces good-tasting dairy products.

A good milking doe goat will produce about 3 quarts each milking, on average, where an average milk cow will produce much more—about three gallons. So your choice will depend on your facilities, labor and needs. Remember that all "extra" milk can be used to produce dairy products such as butter (which can also be used as a cooking fat) and cheese; dairy byproducts can be fed to chickens and hogs. Extra milk can be used to bottle feed young calves or kid goats. On a survival homestead, there is no waste!

When planning on establishing a home kitchen dairy, be sure to stock up on such things as rennet tablets, which make forming cheese curd much easier and more reliable, cheese cultures (as you need for some "fancier" cheeses), cheese cloth and a cheese press or the materials to make one. These materials can be as simple as a #10 can or a 4-inch piece of PVC pipe and wood.

Okay, now we have your family ready with a vegetable garden, small fruits, grains, orchard, meat/eggs, and dairy production. Pretty nifty, right? You bet. For now you can also make soap from used cooking fats, which you can save in a can after each use. Soap making is easy and glitch free, requiring only strained, clean used fat and lye (which can be produced by seeping water down through wood ashes). This soap is great and can be used to wash clothes, babies, and hair.

Add a hive or two of bees, and your sugar requirements are easily met. Then too, the bees will pollinate your entire garden, grain patch and orchard, ensuring bountiful harvests. Bees are easily established and easy to work with. I've only been stung twice working with domestic bees, and probably a few dozen times by "wild" bees.

Survival homesteading is addicting. Once you get started, your mind works constantly at ways you can do more to be less reliant on the system. Now I say "can do" as few homesteaders actually practice every bit of their knowledge. I can raise and shear sheep, spinning wool into yarn to make clothes. And I can tan hides from which to fashion clothes and footwear. But I choose to use my time in other ways, which are more productive to the family at present. But the knowledge is there, should our needs change.

Survival homesteading is rewarding, financially and spiritually, as a basic instinct in human beings is to provide for their own and their family's needs. Never become overwhelmed by feeling that you must do everything at once. It is better to proceed in steady forward steps, rather than to run forward, fall and lose heart. One vital tip: start small and work your way up as your ability and knowledge increases. Your survival homestead depends on it. Δ

Resources

* ***Best of the First Two Years*** published by *Backwoods Home Magazine*
* ***The Third Year*** published by *Backwoods Home Magazine*
* ***The Fourth Year*** published by *Backwoods Home Magazine*
* ***The Fifth Year*** published by *Backwoods Home Magazine*
* ***The Sixth Year*** published by *Backwoods Home Magazine*
* ***The Seventh Year*** published by *Backwoods Home Magazine*
* ***The Eighth Year*** published by *Backwoods Home Magazine*
* ***The Ninth Year*** published by *Backwoods Home Magazine*
* ***The Jackie Clay CD-ROM*** *available from Backwoods Home Magazine*
* ***The Backwoods Home Magazine CD-ROM Years 7-10***
* ***Veterinary Guide for Animal Owners*** by C.E. Spaulding, D.V.M. and Jackie Clay
* ***Cooking with Home Storage*** by Vicki Tate
* ***The Neighborhood Forager*** by Robert K. Henderson
* ***Feeding the Whole Family*** by Cynthia Lair
* ***Mountain Man Crafts and Skills*** by David Montgomery
* ***Native American Crafts*** and Skills by David Montgomery
* ***Country Self-Sufficiency*** by Pete Mickelson
* ***The Encyclopedia of Survival Techniques*** by Alexander Stilwell
* ***Living Well on Practically Nothing*** by Edward H. Romney
* ***The Big Book of Country Living*** by Ernest Thompson Seton
* ***Square Foot Gardening*** by Mel Bartholomew
* ***Saving Seeds*** by Marc Rogers
* ***Small Scale Livestock Farming*** by Carol Ekarius
* ***Achieving Energy Independence One Step at a Time*** by Jeffrey R. Yago
* ***How to Live Without Electricity and Like It*** by Anita Evangelista
* ***In the Gravest Extreme*** by Massad F. Ayoob
* ***The Ayoob Files: The Book*** by Massad F. Ayoob
* ***StressFire*** by Massad F. Ayoob
* ***StressFire II*** by Massad F. Ayoob
Amish Country Cooking by Andy & Millie Yoder
Bread Winners by Mel London
Dehydrated Food Cooking by Stan Smith
Kitchen Magic-Cooking with Common Grains by Deanna Sudweeks
Marlene's Magic with Food Storage by Marlene Peterson
The Natural Nine *(Cooking with the nine most common grains)* by Lorraine D. Tyler

* Denotes this book or CD-ROM is available from *Backwoods Home Magazine.* Call 1-800-835-2418 to order a book or a subscription to the magazine ($21.95 per year)
www.backwoodshome.com